A Discipline-Based Teaching and Learning Center

Gili Marbach-Ad • Laura C. Egan
Katerina V. Thompson

A Discipline-Based Teaching and Learning Center

A Model for Professional Development

Springer

Gili Marbach-Ad
University of Maryland
College Park, MD, USA

Laura C. Egan
University of Maryland
College Park, MD, USA

Katerina V. Thompson
University of Maryland
College Park, MD, USA

Additional material to this book can be downloaded from http://extras.springer.com

ISBN 978-3-319-37783-4   ISBN 978-3-319-01652-8   (eBook)
DOI 10.1007/978-3-319-01652-8

Springer Cham Heidelberg New York Dordrecht London

Printed on acid-free paper

Springer International Publishing AG Switzerland is part of Springer Science+Business Media (www.springer.com)

Foreword

While teaching and learning have always been at the heart of the missions of higher education institutions, considerable interest in recent years pertains to ways to improve learning processes for both undergraduate students and graduate students. Concerns about whether students are being prepared for the lives they will lead, the employment responsibilities they will fill, and their roles in meeting the needs of society come from parents, employers, legislators, foundations, and national agencies, as well as from institutional administrative leaders, faculty members, and students themselves. These concerns have fueled growing interest in initiatives to strengthen undergraduate education, to integrate more preparation for teaching into doctoral education, and to refine curricula in order to deepen student learning outcomes. Encouraging and facilitating learning is a core mission of higher education, and the renewed interest in effectively meeting this mission is resulting in exciting initiatives across many universities and colleges.

While many higher education institutions are re-committing themselves to excellence in teaching and learning processes, the process of engaging in organizational change in support of deeper student learning is not always easy. Good intentions and innovative plans do not always achieve the full results intended. Significant change in teaching and learning processes requires the use of multiple levers for change that recognize the complexity of higher education institutions. For example, institutional leaders at all levels—presidents, provosts, deans, and department chairs—need to play a role in articulating goals, guiding action plans, and supporting faculty in their work to rethink curricula and teaching practices. The academic reward systems, particularly tenure and promotion processes, need to take into account, recognize, and reward the significant effort faculty members must invest to make major changes in their courses and teaching plans. Faculty members, as well as graduate students preparing for teaching roles, need opportunities to learn about the research on learning and on the teaching approaches that are particularly effective in supporting learning in their disciplines and fields. They also need time to integrate what is known about learning and teaching into their own teaching plans and practices. Faculty members also benefit from opportunities to interact and collaborate with colleagues engaged in similar work. Occasions to share successes and discuss challenges, and to engage in collegial conversations that generate new ideas, encourage individual efforts and promote institution-level innovation. To make change in their teaching, faculty members must understand and value the change that is called for and perceive that their efforts will be valued (or at least that they will not be penalized for this work, such as in tenure and promotion decisions)—and, additionally, they must understand and know how to engage in the new approaches to teaching that are being encouraged. In short, reform in teaching and learning processes requires a systemic approach that takes into account the various factors that affect the choices faculty members make about the allocation of their time and work efforts, including how they approach their roles as teachers.

For years, many universities and colleges have helped faculty members grow and improve as teachers through the efforts of institutional faculty development offices or teaching and learning centers. These offices or units have typically offered support through workshops and seminars, consulting for individual faculty members, mid-course evaluations, and, at some institutions, faculty learning communities that convene faculty members with similar interests or learning goals. The efforts of these centers and their leaders have supported faculty members striving to be excellent teachers and provided scaffolding for institutional efforts to improve teaching and learning processes. However, while such centers have done much to advance the teaching missions on their campuses, in recent years there is increasing interest in situating teaching improvement efforts more closely in the disciplinary contexts in which faculty members work. One reason for interest in discipline-based teaching initiatives and professional development is that disciplines have particular cultures that influence how faculty members engage in and organize their work, including their teaching. Furthermore, and of particular relevance to efforts to reform teaching and learning, disciplinary contexts and traditions frame the structure of the knowledge base in fields, how questions are focused and addressed, the relationships between concepts and thus the sequencing of learning processes, and the strategies to guide learning in the discipline. Thus, opportunities to interact with their disciplinary peers, or with those in related disciplines, may be more attractive and useful to faculty members interested in professional development around teaching and learning than exchanges with those situated in more distant disciplinary traditions. Furthermore, in the STEM fields (sciences, engineering, mathematics, and technology) in recent years, a new line of scholarly work has developed called Disciplinary-Based Education Research (DBER). Those prepared as DBER scholars deeply understand their disciplines, while also having thorough knowledge of teaching and learning processes, and particularly of the issues, questions, and challenges pertaining to teaching and learning in their specific disciplines. The increasing number of DBER scholars situated in universities and colleges across the United States is supporting interest in discipline-based efforts, especially in STEM fields, to reform teaching and learning processes at the undergraduate level and, sometimes, at the graduate level.

This book by Gili Marbach-Ad, Laura C. Egan, and Katerina V. Thompson focuses on the very significant and promising trend, gathering momentum across a growing number of institutions, to organize discipline-based teaching and learning centers. In this book, the authors have provided a very readable and thorough examination of one such center that serves faculty members, postdoctoral scholars, and graduate students. The book is organized into six chapters, beginning with a chapter that offers an explanation of and rationale for discipline-based professional development. Chapters 2, 3, 4 and 5 then explain the work, activities, and resources of the particular center that serves as a case study. Chapter 2 discusses how the center uses visiting teachers and scholars as well as seminars and workshops to support discipline-based professional development. Chapter 3 discusses strategies for acculturating new faculty members, while Chapter 4 addresses how the center provides consultation to individuals and groups of faculty. Chapter 5 focuses on strategies particularly organized to prepare graduate students for their responsibilities and roles as teachers. The final chapter makes a compelling case for the importance of evaluating teaching and learning centers, and offers strategies for conducting such evaluation.

Several themes weave throughout the book and enrich its contribution to readers. First, an overarching theme is that change in higher education requires a systemic approach and that universities and colleges should recognize faculty members as essential change leaders in efforts to reform and improve the educational mission. However, to engage in their roles as change leaders, faculty members must have the knowledge and expertise to guide their work. Since preparation for teaching responsibilities has not traditionally been part of academic preparation, current faculty members need specific and systematic support. Future faculty (today's graduate students) also benefit from opportunities to develop as teachers, as part of the graduate experience.

Second, preparation and professional development focused on teaching should be supported by theory, literature, and research. Much is known (and the research is expanding continuously) about learning processes, as well as about effective teaching. Discipline-based education research (DBER) is adding important insights into learning and teaching within specific disciplinary contexts. This book effectively draws on these bodies of research and literature and shows how such research can directly inform and guide the strategies used by a discipline-based teaching and learning center. The authors use the concept of pedagogical content knowledge (PCK)—attention to the intersection of what is taught (content), how it is taught (pedagogy), and the context in which it is taught—as the guiding framework for the book. Throughout the volume, they highlight five components of PCK (i.e., knowledge of student understanding of science, knowledge of science curriculum, knowledge of instructional strategies, knowledge of assessment of student learning, and orientation to science teaching) and effectively use these components as the organizing structure for their chapters. This approach emphasizes consistently that the design and work of a teaching and learning center is most effective if the center is informed by research and theory.

The third theme that is highlighted throughout each chapter is that the improvement of teaching and learning requires carefully planned practical interventions and strategies. Building on theory, research, and a clear conceptual framework, the authors allocate half of each chapter to extensive explanation and rationale for each professional development strategy they recommend. These detailed descriptions of the creative programs, resources, and services they offer in their center provide an excellent template to inspire the work of those starting similar centers. An implied message is that those engaged in designing professional development for faculty members and graduate students are wise to recognize that careful planning ensures that those participating will find their time was used efficiently and effectively, and that the investment was worthwhile. Such attention to the practicalities of planning innovative and effective programs is evident throughout this volume.

Finally, Marbach-Ad, Egan, and Thompson show that teaching, like research, is a scholarly activity that deserves serious, sustained intellectual engagement. Their goal, they explain, is to help current and future faculty members shape identities as what they call "reflective, reform-minded teachers." In this well-argued volume, informed by theory and research and attentive to the importance of carefully designed practice, they show how discipline-based professional development can serve as a powerful and effective lever for guiding faculty members toward identities that include commitment to effective and reflective teaching. The map they provide for creating discipline-based professional development should inspire collaborative efforts among colleagues at other institutions—including institutional administrative leaders, discipline-based education researchers, and faculty developers, as well as faculty members and graduate students—to design innovative approaches to improve teaching and learning. The students we all serve will benefit.

Professor, Higher, Adult, and Lifelong Education   Ann E. Austin
Michigan State University
East Lansing, MI

Acknowledgment

This book, and even more so the success of the Teaching and Learning Center (TLC), is the product of the efforts of many people. While we cannot list everyone, we would like to thank all of the people who have contributed to the TLC and to the writing of this book.

First and foremost, we are deeply grateful to the undergraduate students, graduate students, postdoctoral fellows, and faculty members of the University of Maryland. They have inspired our work and been key collaborators throughout. Much of what we share in this book grew out of their ideas, feedback, and willingness to participate in our professional development and research activities. We would also like to thank the leadership of the College of Chemical and Life Sciences and the College of Computer, Mathematical and Natural Sciences, particularly **Norma Allewell, Robert Infantino, Jayanth Banavar, Joelle Presson,** and the current and former department chairs **Norma Andrews, Michael Doyle, William Fagan, Charles Mitter, Janice Reutt-Robey,** and **Gerald Wilkinson**. They played a fundamental role in the creation of the TLC, and have offered unwavering support for the Center and its mission of improving undergraduate instruction. We offer a special thanks to the many people from across the university with a longstanding commitment to teaching and learning. **Spencer Benson**, the former director of the University of Maryland Center for Teaching Excellence, was a strong supporter during the creation of the TLC and has been a frequent collaborator who was always willing to share his expertise and vision. We would also like to thank the broader science education research community, particularly the Visiting Teacher/Scholars who shared their expertise with our college community.

Additionally, we would like to thank the following individuals:

Katie Schaefer Ziemer, whose work as a graduate assistant from the creation of the TLC greatly enriched the Center's research and evaluation program

Michal Orgler, who added a valuable psychological perspective to the TLC's evaluation processes and instruments

Ann Smith, who holds great expertise in teaching, science education, and leadership, and who has tirelessly pushed us in pursuit of our mission

Virginia Anderson and **Charles Henderson**, who served as external evaluators of the TLC and provided insightful feedback that contributed to refining the TLC's professional development and evaluation activities

Judy Dori, whose thorough review and thoughtful comments added to the clarity and coherence of this book

Mike Landavere, Mel Manela, and **Chris Camacho**, who patiently helped us overcome all of the technological obstacles we encountered

Loretta Kuo, whose graphic design adds clarity to the text

Omer Ad and **Michal Ad**, whose drawings, figures, and photos liven up the pages of this book

Ann Austin, who graciously contributed the foreword for this book

Bernadette Ohmer and **Marianna Pascale**, whose advice and support have made the book possible

Lastly, we would like to offer our thanks to our families. Their continual encouragement deepened our dedication to our work. They have lovingly supported us as we worked around the clock to do the work that we are so passionate about.

The TLC was supported by a grant from the Howard Hughes Medical Institute Undergraduate Science Education Program to the University of Maryland and by a Course, Curriculum, and Laboratory Improvement grant from the National Science Foundation (DUE–0942020).

Contents

About the Authors

Gili Marbach-Ad is a Research Associate Professor and the Director of the Teaching and Learning Center in the College of Computer, Mathematical, and Natural Sciences at the University of Maryland. She holds B.S. and M.S. degrees in Biology from Tel Aviv University. She received her Ph.D. in Science Education in 1997 from Tel Aviv University; the subject of her dissertation was "Students' Conceptions in Genetics." She has been involved in numerous aspects of science education, including the development of study tools (science curricula, learning materials, computer software), and the training of pre- and in-service teachers through courses she has taught at Tel Aviv University. From 2000 to 2006, Marbach-Ad was a faculty member, head of the Life-Science Education Division with a tenure position in the Department of Science Education at Tel Aviv University. She joined the University of Maryland faculty in 2004. Currently, her research focuses on three topics: defining student conceptions and assessing curricular and pedagogical reforms; implementing active-learning approaches in science courses; and building a model for teacher preparation programs. Marbach-Ad publishes regularly in top peer-reviewed journals in the field. (For more biographical information, go to cmns-tlc.umd.edu/tlc/gilimarbachad.)

Laura C. Egan is a doctoral student in Teaching and Learning, Policy and Leadership, with a specialization in Education Policy, at the University of Maryland. As a doctoral student, she has served as graduate research assistant in the Teaching and Learning Center in the College of Computer, Mathematical, and Natural Sciences. Prior to coming to the University of Maryland, Laura earned a B.A. in Political Science and Spanish from Wake Forest University and an M.A. in Development Management and Policy from the University of San Martín in Buenos Aires. Laura has worked in research and evaluation positions at all levels of the educational system. Currently, she is a Research Associate at Westat, where she is Project Director for an education study. Laura's research interests focus on how research can inform education policy, and the relationship between policy design, policy implementation, and educational outcomes.

Katerina V. Thompson is Director of Undergraduate Research and Internship Programs in the College of Computer, Mathematical, and Natural Sciences at the University of Maryland. She holds B.S. and M.S. degrees in Biology from Virginia Tech and a Ph.D. in Zoology from the University of Maryland. She is also a Smithsonian Institution Research Associate in the Department of Reproductive Sciences, National Zoological Park, where her research interests focus on social influences on reproductive behavior and physiology. In addition to facilitating student involvement in co-curricular experiences, she coordinates externally funded curriculum development initiatives in the biological sciences and oversees the CMNS Teaching and Learning Center, which provides professional development opportunities for science faculty and graduate students. She has served as PI or Co-PI on science education grants from the Howard Hughes Medial Institute (HHMI) and the National Science Foundation (NSF). Her science education research focuses on strategies for facilitating student success and teaching reform in higher education.

Chapter 1
Discipline-Based Professional Development

*My real interest in professional development in teaching is
ultimately to find new and better ways to engage my students
and to help them learn and do better.*
    *The CLFS Teaching and Learning Center has supported my
work in improving my teaching. I work closely with the TLC on
projects, and have benefited from invited speakers and
workshops. I have also had the opportunity to work with faculty
in teaching communities ... Discussions in these communities
have supported and motivated my interest in teaching. I like
being involved in a common mission.*

–Faculty members reflecting on the importance of
professional development in teaching and the role of the TLC

Higher education is undergoing a major transformation that encompasses both
philosophy and practice. Where it was once believed that subject matter expertise
alone was a sufficient basis for teaching at the university level, there is now a
growing recognition that teaching expertise is also necessary to ensure that students
develop a deep, lasting understanding of that subject matter. As a result, institutions
of higher education have placed increasing emphasis on professional development
to help university faculty members compensate for their lack of specific training
in teaching. This book describes one such response—a university teaching and
learning center based within a disciplinary unit and focused specifically on teaching
within interrelated disciplines.

In 2006, the College of Chemical and Life Sciences (CLFS) at the University
of Maryland established the Teaching and Learning Center (TLC) to support its
faculty members, postdoctoral fellows (postdocs), and graduate students through
professional development programming. The TLC was similar in organization and
programming to the campus-wide teaching center in existence at that time; however,
it differed in scope in that it focused on a limited range of interrelated disciplines.
This well-defined disciplinary emphasis has provided a distinct advantage in that
the activities of the TLC are deeply integrated into the missions of the CLFS
departments. Over the years, our College community has come to value the
TLC's workshops and seminars, special programming for new instructors and
graduate students, and consulting services. In the pages that follow, we present the

© Springer International Publishing Switzerland 2015   1
G. Marbach-Ad et al., *A Discipline-Based Teaching and Learning Center*,
DOI 10.1007/978-3-319-01652-8_1

discipline-based teaching and learning center as an effective model for professional development for university-level educators as they embark in the challenging task of transforming science education.

Transforming Undergraduate Science Education

Undergraduate education is in the midst of a major transformation to improve its quality and relevance. The transformation stems from widely acknowledged deficiencies in science, technology, engineering, and mathematics (STEM) education (Arum & Roksa, 2011; Austin, 2011; President's Council of Advisors on Science and Technology (PCAST), 2012). Currently, less than 40 % of undergraduate students who intend to major in a STEM field complete a STEM degree (Gates & Mirkin, 2012; PCAST, 2012). Students who have left STEM majors often cite the uninspiring introductory courses and an unwelcoming, unsupportive environment as contributing to their decision to leave these fields (Seymour & Hewitt, 1997). Moreover, many of the students who do graduate with a STEM degree lack the scientific knowledge and skills to compete in the global market (National Research Council (NRC), 2012). These deficiencies threaten the quantity and quality of future scientists as well as future entrepreneurs, educators, and policymakers, all of whose work requires scientific literacy. One way to boost STEM student persistence and preparation is to improve STEM education (Association of American Universities (AAU), 2011; NRC, 2012; PCAST, 2012).

We see university faculty members as key players in efforts to improve STEM education because of their position as front-line educators. In large research universities, graduate students also fill this role in their capacities as teaching assistants. However, neither faculty members nor graduate students receive extensive training in teaching. Faculty members usually come to the university well prepared to assume their roles as scientific researchers, but they tend to be less prepared for their teaching responsibilities (Fairweather, 1996, 2008; Massy, Wilger, & Colbeck, 1994). Similarly, graduate students receive extensive training in conducting scientific research but rarely receive adequate professional development to hone their teaching skills (Austin, 2002; Golde & Dore, 2001). Absent this training, both graduate students and faculty members generally base their teaching approaches on their own experiences as students, in which lecture-based teaching approaches typically predominated. As a result, traditional teaching approaches have proven to be very persistent in spite of the growing body of research supporting the use of teaching approaches that engage students in active learning (PCAST, 2012).

Active Learning and Evidence-Based Teaching Approaches

Evidence-based teaching approaches are rooted in learning theories that suggest that students learn better when they actively construct their knowledge rather than receive it passively (Ausubel, 1968; Bruner, 1960; Dewey, 1897; Piaget, 1954). Teaching approaches that support the active construction of knowledge promote the development of critical thinking and higher-level learning (Wieman, 2007). While much of the early research on the effectiveness of active learning approaches was based in K-12 education, an expanding body of research similarly demonstrates that these approaches also enhance student learning at the post-secondary level (Dori & Belcher, 2005; Freeman, Haak, & Wenderoth, 2011; Freeman et al., 2007; Injaian, Smith, German Shipley, Marbach-Ad, & Fredericksen, 2011; Jensen & Lawson, 2011; Kitchen, Bell, Reeve, Sudweeks, & Bradshaw, 2003; Knight & Wood, 2005; Senkevitch, Marbach-Ad, Smith, & Song, 2011; Smith, Wood, Krauter, & Knight, 2011; Udovic, Morris, Dickman, Postlethwait, & Wetherwax, 2002; Walker, Cotner, Baepler, & Decker, 2008). A recent meta-analysis of 225 studies demonstrated that undergraduate students perform better with active learning techniques than with traditional lecture across STEM disciplines (Freeman et al., 2014). Active learning approaches have been demonstrated to be effective in a variety of different university settings—introductory and advanced courses, for majors and non-majors, for low and high performing students—using robust evaluation techniques, as the following examples illustrate.

Reducing Failure Rates and Increasing Student Learning in Introductory Courses Freeman et al. (2011) compared student learning outcomes in redesigned and traditional introductory biology course for majors. The redesigned course included six key components: Socratic lecturing; regular reading quizzes; in-class ungraded active-learning exercises, such as clicker questions; in-class group discussions of exam-style questions; frequent online peer-graded practice exams; and graded weekly class note summaries. In the redesigned course, the authors found that failure rates declined by approximately 66 % compared to the traditional course and students earned better scores on exams emphasizing questions that were classified as high-level according to Bloom's taxonomy (Bloom, 1984).

Increasing Scientific Reasoning in an Upper-Level Course Kitchen et al. (2003) examined the effectiveness of a large-enrollment cellular biology course in helping biology majors acquire skill in interpreting experimental data. The authors compared a traditional, lecture-based course to a redesigned version that used a workshop format and active learning approaches. The redesigned course resulted in significant increases in the students' scientific reasoning and ability to draw conclusions from experimental data compared to the traditional lecture-based course. Students found this new approach to be challenging, but generally endorsed this approach to learning and the analytic emphasis. Course instructors also indicated that, compared to assessments based on recall, assessments based on data analysis provided them with better information to evaluate students' performance and provide students with additional support.

Combining Instructor-Centered and Learner-Centered Approaches Smith et al. (2011) compared student learning in genetics courses that used one of three approaches: an instructor-centered approach; a learner-centered, peer discussion constructivist approach; and a blend of the two that consisted of lecture following peer discussion activities. Using a concept inventory, they found that majors and non-majors of all ability levels (weak, medium, and strong performers, based on success in clicker responses over the semester) learned the most when the combination approach was used, which indicates that the different approaches may have synergistic effects. The difference in learning gains between the instructor-centered and student-centered approaches was greatest for the top-performing students, who benefitted little from the instructor-centered approaches.

Despite strong evidence regarding the superiority of active learning over passive learning, there are multiple barriers to the widespread adoption of active learning approaches in the university classroom. First, many faculty members are unfamiliar with active-learning methods and may not be aware of the research indicating their impact on student learning (PCAST, 2012). Furthermore, even faculty members who are aware of the value of active learning approaches are rarely trained in how to implement these approaches effectively in their classrooms (PCAST, 2012). This lack of training is particularly important, as poor fidelity in implementing specific approaches can impede student learning (Andrews, Leonard, Colgrove, & Kalinowski, 2011; Henderson & Dancy, 2008) and may cause faculty members to become skeptical of the value of active-learning approaches generally. Professional development programs can address these issues.

Professional Development for Teaching in the University

Universities and academic societies have long offered professional development programming, which tends to take one of three forms:

1. Universities commonly offer discrete, a la carte workshops and seminars through campus-wide teaching and learning centers that serve all disciplines.
2. Both universities and professional societies sometimes sponsor intensive, multi-day programs (e.g., summer teaching institutes) that focus on specific teaching strategies, philosophies, or subject matter.
3. Universities and professional societies may offer fellowship programs, often based on a competitive selection process, or support faculty learning communities (FLCs) that foster extended interaction and engagement.

All three approaches have strengths as well as shortcomings. Campus-level programs are often perceived as irrelevant because their content is general rather than discipline-specific. One-time seminars and intensive teaching institutes may help increase awareness of effective approaches, but often lack sustained support for their implementation. Fellowships and other extended special programs provide stronger support for participants, but usually serve only a small percentage of a university's teaching community.

In recent years, a number of universities have created STEM-specific education centers to support undergraduate teaching and learning in the sciences. The STEM Ed Center project has compiled a list of more than 50 such STEM education centers (serc.carleton.edu/StemEdCenters/profiles.html). These centers have diverse missions, audiences, and mechanisms for improving STEM education. Although some STEM education centers are housed in a STEM college or department, most are campus-wide entities.

Our Teaching and Learning Center is an example of a STEM-specific education center housed in a STEM college. We take a highly specialized approach in that we focus on providing professional development to chemistry and biology faculty members, postdocs, and graduate students. This discipline-specific approach stems from our belief that professional development activities should integrate best teaching practices with the content being taught and the context in which it is taught. By focusing on chemistry and biology, we provide professional development that integrates pedagogy and content in a way that is highly relevant to our audience. This approach is built upon the theory of pedagogical content knowledge.

Pedagogical Content Knowledge

Shulman (1986) coined the term pedagogical content knowledge (PCK) to represent the integration of what we teach (content) and how we teach (pedagogy) into a comprehensive knowledge base for teaching (Grossman, 1990; Shulman, 1986, 1987). Grossman (1990) suggested that PCK integrates not only content knowledge and pedagogical knowledge but also knowledge of context. Figure 1.1 provides a visualization of Grossman's model of teacher knowledge. Content knowledge

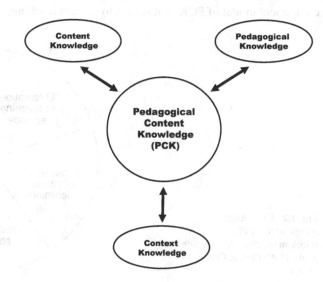

Fig. 1.1 Pedagogical Content Knowledge (PCK) and its inter-relatedness with content, pedagogical, and context knowledge

refers to subject area knowledge. Pedagogical knowledge includes components such as classroom management, curriculum, assessment, and instructional techniques. Context refers to factors that impact student learning, such as type of school, classroom environment, and student characteristics. PCK is a unique domain of knowledge that results from the transformation of content knowledge through the use of appropriate pedagogy, given the context in which the teaching and learning are occurring, to create forms that students can understand (Geddis, Onslow, Beynon, & Oesch, 1993; Grossman, 1990; Magnusson, Krajcik, & Borko, 1999; Park & Oliver, 2008; Shulman, 1986).

PCK has gained prominence as a conceptual tool, focus of research, and framework for providing professional development for current and future teachers. Teachers develop PCK through training as well as through practice and reflection (Eraut, 1994; Park & Oliver, 2008). PCK, content knowledge, pedagogical knowledge, and context knowledge share a reciprocal relationship in that growth in one domain can result in growth in another domain. PCK theory suggests that teacher professional development is most effective when knowledge bases are taught in a purposefully integrated manner that reflects best practices in education (Gess-Newsome, 1999).

The Five PCK Components

Over the years, many scholars have further refined and/or modified the conceptualization of PCK (Cochran, DeRuiter, & King, 1993; Fernandez-Balboa & Stiehl, 1995; Geddis et al., 1993; Grossman, 1990; Hashweh, 2005; Injaian et al., 2011; Loughran, Berry, & Mulhall, 2006; Magnusson et al., 1999; Smith & Neale, 1989; Tamir, 1988; van Driel, Verloop, & De Vos, 1998). In this book, we use a five-component model of PCK as it relates to science teaching, shown in Fig. 1.2.

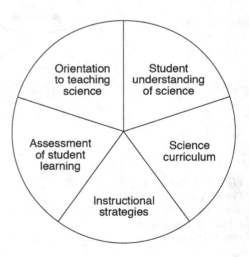

Fig. 1.2 Knowledge components of PCK (Grossman, 1990; Magnusson et al., 1999; Park & Oliver, 2008)

The five PCK components are fundamental to the work of educators both inside and outside of the classroom. These components are interrelated and can overlap. Below, we provide examples of what each PCK component encompasses.

1. **Knowledge of student understanding of science** refers to understanding the prior knowledge that students have, as well as their alternative conceptions that can impede their understanding. This component also addresses student diversity in terms of their prior knowledge.
2. **Knowledge of science curriculum** addresses the identification of learning goals for a specific subject, and determining how to achieve those learning goals through vertical and horizontal curricula. This PCK component includes how to select and sequence content within a course and across courses to build from basic to deep understanding.
3. **Knowledge of instructional strategies** addresses the teaching methods appropriate for a given subject matter and level of complexity. Knowledge in this area includes awareness of evidence-based instructional strategies and how to apply them.
4. **Knowledge of assessment of student learning** refers to understanding what is important to assess and how to assess it. This includes how to develop and implement different types of formative and summative assessments to get comprehensive feedback on student learning.
5. **Orientation to science teaching** refers to educator knowledge and beliefs about the way science should be taught and the goals of teaching science. This orientation is content and context specific, and shapes the way that a teacher approaches teaching. In recent years, the knowledge and beliefs about how science should be taught have drawn heavily from the scientific teaching approach (Handelsman et al., 2004; Handelsman, Miller, & Pfund, 2007).

Scientific Teaching
Scientific teaching refers to teaching science in a way that reflects the processes and rigor associated with scientific research (Handelsman et al., 2004, 2007; NRC, 2012). This approach draws on the recommendations of backwards design (Wiggins & McTighe, 1998) by first setting learning goals, then determining what would demonstrate the achievement of these goals, and only then designing activities to allow students to achieve the desired learning outcomes. Scientific teaching takes an explicitly scientific approach to backwards design and requires iterative evaluation of the methods used through assessment of student learning outcomes (Handelsman et al., 2007). Scientific teaching leverages the culture of science to improve science education and is particularly appropriate for science educators at the undergraduate level, who are trained as scientists.

PCK for the Undergraduate Level

Much of the existing research on PCK focuses on instruction at the primary and secondary levels (Lederman & Gess-Newsome, 1999). Instruction at the postsecondary level differs not just in terms of its context and complexity, but also in the professional roles of the instructors. Fernandez-Balboa and Stiehl (1995) elaborated on the key differences between instruction at the K-12 and postsecondary levels as follows:

- University classes vary in size more than primary and secondary classes, with some university classes—including many in the sciences—being an order of magnitude larger than K-12 classes;
- Undergraduate students are more mature and have made a choice to continue their education, and as a result may be more engaged in their learning;
- The content taught at the undergraduate level tends to be more complex and specialized than content in lower levels; and
- Educator preparation at the undergraduate level focuses on mastery of the content area, whereas K-12 teacher preparation places more emphasis on pedagogy.

Professional development programs for instructors should reflect the context in which they are teaching. We have found that a discipline-based professional development program is ideally positioned to help faculty members overcome their limited pedagogical training and develop their capacity to transform their extensive content knowledge into a form that is understandable to their students. This disciplinary focus allows us to provide carefully tailored professional development with an emphasis on pedagogical content knowledge.

Overview and Brief History of the Teaching and Learning Center (TLC)

Our Teaching and Learning Center was created in 2006 to serve the College of Chemical and Life Sciences at the University of Maryland, College Park. The College of Chemical and Life Sciences (CLFS) was comprised of four departments: biology, entomology, cell biology and molecular genetics, and chemistry and biochemistry.[1] The University of Maryland is classified by the Carnegie Foundation as a research university (very high research activity). At the time of publication, it enrolled more than 25,000 undergraduate and 9,500 graduate students in approximately 110 undergraduate and 90 graduate programs. Within the chemical and life

---

[1]In 2010, the College of Chemical and Life Sciences merged with the College of Computer, Mathematical, and Physical Sciences to become the College of Computer, Mathematical, and Natural Sciences (CMNS). The TLC continues to serve the four biology and chemistry departments at present, but is beginning to scale up its services to the entire CMNS.

sciences departments, there were approximately 300 faculty members,[2] about 2,200 undergraduates pursuing majors in these departments, and 400 graduate students. The scale of our university and our college suggested that a concerted, coordinated effort was needed to provide professional development to our large population of faculty members and graduate students. Therefore, we developed a disciplinary TLC to serve the specific needs of chemistry and biology faculty members, based on the premise that faculty members' disciplines tend to shape their teaching behavior more than the norms of and connections to their institution (Fairweather, 1996). The disciplinary focus of our center makes it a powerful model for developing and supporting pedagogical content knowledge and scientific teaching.

One impetus for the creation of the center was an external review of the College's funded science education and curriculum development programs. The review process, which involved groups of faculty members and graduate students who were engaged in curriculum enhancement projects, highlighted a need for more professional development focused on science instruction. Our faculty members indicated that they were largely unaware of national STEM education reform efforts that would be complementary to the goals of their curriculum enhancement projects. Likewise, our graduate students indicated that they often felt unprepared to assist faculty members in redesigning or revising courses. The CLFS, in coordination with the campus-wide Center for Teaching Excellence, created the TLC to provide structured, focused professional development opportunities for chemistry and biology faculty members and graduate students. Initial funding for the establishment of the TLC came from an Undergraduate Science Education Program grant from the Howard Hughes Medical Institute and a Course, Curriculum and Laboratory Innovation grant from the National Science Foundation.

The TLC has three overarching goals: (1) to provide opportunities for science faculty members, postdocs, and graduate students to consult and collaborate with science education experts, (2) to make training in teaching science part of the standard graduate program alongside training in scientific research, and (3) to create a structured environment of teaching and learning communities that supports efforts to improve teaching. To accomplish these goals, the TLC provides a wide variety of resources to faculty members, postdocs, and graduate students: seminars and workshops, acculturation for new faculty members, programs for graduate students, consultation for individuals, and consultations for groups of faculty (Fig. 1.3).

Key Contributors to a Disciplinary TLC

A disciplinary teaching and learning center is in a unique position to foster collaboration between scientists and educators. STEM education reform requires

---

[2]By faculty members, we refer to full-time and part-time faculty members, including both non-tenure-track instructional and tenured/tenure-track faculty.

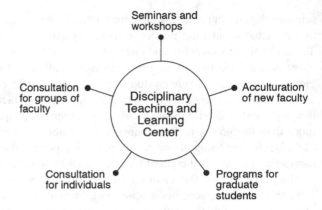

the integration of science and education, but these disciplines are traditionally housed in different departments or administrative units with limited avenues for interaction. Henderson, Beach, and Finkelstein (2011) identified at least three distinct research communities that have a role in post-secondary STEM education reform and professional development. The first group is *STEM education researchers*, who are typically based in STEM departments and focus on student learning within STEM disciplines. The second community is comprised of *faculty development researchers*, who are generally associated with campus-wide centers for teaching and learning and provide professional development for faculty members from all disciplines across the institution. The third community, *higher education researchers*, tends to reside in departments of educational leadership within colleges of education or in university administration, and generally focus on how cultural norms, organizational structures, and governmental policy influence higher education practices.

While these three communities often act in discrete arenas, they serve complimentary roles. Consequently, professional development can be enhanced when it includes all three research communities. Furthermore, given the key role that science faculty members and graduate students play as frontline educators, they also have integral roles in developing, implementing, and evaluating STEM education reform initiatives.

The founding director of the TLC is a science educator with an M.S. in Biology, a Ph.D. in Science Education, and extensive experience in faculty professional development. She participates in all three research communities, which allows her to serve as a bridge between the disciplines of science and education. Additional support for TLC programming and research is provided by a graduate assistant, who typically has been pursuing an advanced degree in a discipline complimentary to the mission of the TLC (e.g., Psychological Measurement, Education).

Importantly, the TLC relies heavily on the collaboration of science education researchers from the biology and chemistry departments, who include both tenure-track and non-tenure track faculty members, as well as department administrators. For example, we collaborate with several faculty members who began their professional careers as science researchers but have expanded or shifted their research focus to encompass science education. Some of these faculty members also supervise graduate students who are pursuing science education doctorates in collaboration with the College of Education. Non-tenure-track instructors who have assumed major roles in leading course redesign efforts also play an integral collaborative role with the TLC. Additionally, the TLC relies on the engagement of higher education specialists from across the university as well as other universities, most of whom do not have specific emphasis on science education. Some of our collaborators have been involved with the TLC on an ongoing basis, while others participate in specific projects or initiatives over a more limited time period.

The different groups that collaborate with the TLC each have their own norms and arenas for professional communication (Henderson et al., 2011). Different groups, such as science researchers, science education specialists, and higher education specialists, tend to work independently and to disseminate their research in unique venues. A discipline-based TLC can both connect collaborators to new avenues for professional communication and help bridge differences between groups in terms of research focus, style, and terminology to foster richer collaboration across these actors. By engaging all of these communities in collaborative efforts, the TLC provides a mechanism for the exchange of ideas that otherwise would not cross research community boundaries.

Institutional Levels and External Entities Involved in Change Efforts

The key contributors to a disciplinary teaching and learning center are potential change agents for reforming science education. Science educators—faculty members, postdocs, and graduate students—are ultimately responsible for changing how science is taught, but they cannot do this alone. Their work is impacted by leadership at multiple levels of the institution: department leadership (e.g., chair, directors of graduate and undergraduate programs), college administration (e.g., deans), and university administration (e.g., president, provost, deans of undergraduate studies). A variety of external entities also impact the change process through their interactions with science educators and leadership at multiple levels of the institution. These entities include governmental agencies, professional societies, policy organizations, and external funders (e.g., U.S. Department of Education, American Association for the Advancement of Science [AAAS], Association of Public and Land-Grant Universities [APLU], President's Council of Advisors on Science and Technology [PCAST], National Science Foundation [NSF], and the Howard Hughes Medical Institute [HHMI]).

The push for change can come from any one of these institutional levels or external entities, or from multiple levels or entities simultaneously. For example, a provost may initiate a strategic plan—perhaps motivated by external policy recommendations or accreditation board findings—that includes a campus-wide call for teaching innovation. This plan could then spur deans and department chairs to enact changes within their respective units. Alternatively, a faculty member may initiate change in his or her own course, which then stimulates changes in related courses within a given curriculum. Another common impetus for change is when an external entity provides funding to incentivize a specified change initiative, such as through funding a postdoc to develop a new course in an emerging field. Regardless of where they begin, change efforts have the greatest chance of success when addressed systematically, consistently, and in a way that impacts a large number of faculty members (Austin, 2011; Wieman et al., 2010).

Typically, departments play an important role in change efforts, especially in research universities that employ large numbers of highly specialized faculty members (Wieman et al., 2010). Departments generally make key instructional decisions, including what is taught and how it is taught, as well as decisions related to faculty training opportunities, workload, and performance incentives (Austin, 1994, 1996, 2011; Wieman et al., 2010). They also enjoy a scale and degree of professional cohesion that makes them an appropriate place for engendering bottom-up and top-down support for changes. Departmental leadership can create cultures in which teaching excellence is valued, as well as venues and structures that facilitate innovation in teaching (Austin, 2011). Even when decisions are made at higher levels of the organization, departments often serve as a "linking pin" through which departmental leadership translate institutional messages and connect institutional priorities with faculty work (Bensimon, Ward, & Sanders, 2000; Chu, 2006; Leaming, 1998).

A discipline-based teaching and learning center that is closely tied to the department(s) with which it works can facilitate change by working with key contributors and connecting change initiatives that stem from different institutional levels and external entities (see Fig. 1.4).

Credibility and Buy-In

For a teaching and learning center to serve the role of change facilitator and, at times, as a change agent itself, it must have credibility and buy-in from both science educators and academic leadership. The TLC emerged as a result of our college and departmental leadership's commitment to improving science education. This commitment grew from broad support for this goal at all institutional levels. College leadership has presented a consistent message about the importance of teaching and learning and the value of the TLC in promoting this goal. This leadership buy-in fostered a supportive environment for faculty members, postdocs, and graduate students to pursue professional development in teaching. Without such support, some potential participants might have worried that involvement in professional development activities could brand them as inadequate teachers or stigmatize them as valuing teaching above research. Leadership was also willing to provide

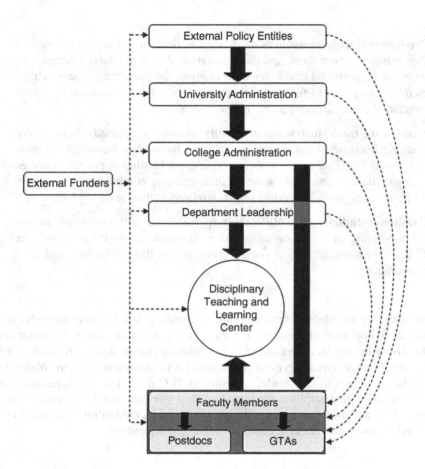

Fig. 1.4 The institutional levels and external entities involved in change efforts

faculty members with release from some professional responsibilities to enable them to invest time in implementing and assessing teaching innovations. Finally, the college provided substantial material support. Notably, our college provided office space for the TLC in the college administration building, a location that highlights the integration of the TLC with administrative priorities. The college leadership also committed to sustaining the TLC beyond the initial period of external funding.

The TLC has worked to build the buy-in of our constituents by being very sensitive to the unique position of each population we seek to serve and designing our activities accordingly, as examples in the text box illustrate. We have found that our populations appreciate this sensitivity and individualization of services, and this has led to a high degree of commitment to the TLC mission and programming on the part of participants.

Pre-tenure faculty members face a great deal of pressure to establish themselves as researchers, and they have limited time available to engage in intensive professional development to improve their teaching. These faculty members may benefit from workshops or support that is individualized and targeted at concrete, manageable interventions.

Non-tenure-track instructional faculty members generally have heavy teaching loads and limited time available to focus on educational research endeavors. The TLC can help them obtain grant funding to provide summer salary or release time from some of their teaching commitments, which in turn can enable them to conduct research related to their teaching initiatives.

Graduate students have different responsibilities and workloads across the timeframe of their graduate studies. Because of these variations, the TLC offers graduate student programming that is flexible in pace and time commitment.

Another key to establishing credibility is engaging faculty members who are respected within their department for their strong research and/or instructional skills. These faculty members can play key roles as change agents by connecting like-minded individuals and promoting avenues for implementing reform. We invite these change agents to take leadership roles in TLC activities, which results in a synergistic effect that amplifies the impact of their work and the TLC's work. They serve as TLC ambassadors amongst their peers, and their credibility gives weight to their endorsement that the TLC could not achieve independently.

How to Initiate a Disciplinary TLC

A discipline-based teaching and learning center may be situated within a single department, multiple related departments, or a college. Creating and sustaining a disciplinary teaching and learning center requires an understanding of the target audience's goals, needs, and existing initiatives, as well as available resources. We recommend conducting a comprehensive needs assessment as an initial step in creating a teaching and learning center. The needs assessment should define needs, problems, assets, opportunities, and purposes (Stufflebeam, Madaus, & Kellaghan, 2000). As a product of our needs assessment process, we created profiles of each department's existing teaching and learning initiatives and related professional development activities, as well as opportunities for growth in these areas. Our needs assessment model, process, and products are described in more detail in Chap. 6. Our needs assessment informed the development of the comprehensive program of professional development that is described in this book.

Outline of the Book

This book offers a theoretical and practical guide to implementing a discipline-based teaching and learning center to provide professional development to faculty members, postdocs, and graduate students. Each chapter of the book highlights one major facet of the TLC's professional development activities and is subdivided into two sections. The first section provides an overview of the theory and research supporting this type of activity. The second section of each chapter is an implementation guide with materials and practical tips for implementing program components and activities. The five PCK components serve as an organizational framework for the book, and we describe activities in terms of the PCK component they address.

Chapter 2 presents our program of workshops and seminars for faculty members, postdocs, and graduate students. These workshops and seminars introduce participants to salient topics in science education and engage them in conversation about how to implement innovative approaches in their own classes. This ongoing programming serves as a gateway to the TLC for new participants and provides a structured environment for interaction around teaching and learning.

Chapter 3 addresses the acculturation of new faculty members, who come with diverse backgrounds and experiences. We conducted a longitudinal study to better understand the needs of new faculty members and which types of professional development programming they used. Our current acculturation programs include group and individual offerings to support new faculty members as they assume their teaching responsibilities and integrate into our college community.

Chapter 4 describes our personalized consultations, which cover a wide variety of topics related to teaching and learning. We offer consulting services of two types: for individuals and for groups of faculty members. These consultations help faculty members implement curricular and instructional innovations, evaluate the impact of these innovations, and obtain grant funding in support of their teaching efforts. Many of our group consultations occur within the context of faculty learning communities (FLCs). These FLCs provide long-term opportunities for faculty members to engage in improving their teaching and collaborate on shared enterprises of greater scope than individual efforts.

Chapter 5 outlines the professional development activities aimed at supporting graduate students in their role as teaching assistants and preparing them to be the faculty of tomorrow. Our programming includes a mandatory preparatory course for all new graduate teaching assistants, an optional course on science teaching and learning in higher education, and a certificate program for graduate students with an interest in gaining further expertise in teaching and learning.

Chapter 6 summarizes the TLC's comprehensive evaluation process. This evaluation includes needs assessment and periodic evaluations of the effectiveness of the resources and support that we provide, using both qualitative and quantitative approaches. Our evaluation measures five levels of program outcomes: participation, satisfaction, learning, application, and impact.

References

Andrews, T. M., Leonard, M. J., Colgrove, C. A., & Kalinowski, S. T. (2011). Active learning not associated with student learning in a random sample of college biology courses. *CBE Life Sciences Education, 10*, 394–405.

Arum, R., & Roksa, J. (2011). *Academically adrift: Limited learning on college campuses.* Chicago, IL: University of Chicago Press.

Association of American Universities (AAU). (2011). Five year initiative for improving undergraduate STEM education. http://www.aau.edu/WorkArea/DownloadAsset.aspx?id=12590

Austin, A. E. (1994). Understanding and assessing faculty cultures and climates. In M. K. Kinnick (Ed.), *Providing useful information for deans and department chairs* (New Directions for Institutional Research, pp. 47–63). San Francisco, CA: Jossey-Bass.

Austin, A. E. (1996). Institutional and departmental cultures and the relationship between teaching and research. In J. Braxton (Ed.), *Faculty teaching and research: Is there a conflict?* (New Directions for Institutional Research, pp. 57–66). San Francisco, CA: Jossey-Bass.

Austin, A. E. (2002). Preparing the next generation of faculty: Graduate school as socialization to the academic career. *The Journal of Higher Education, 73*(1), 94–122.

Austin, A. E. (2011). *Promoting evidence-based change in undergraduate science education.* A paper commissioned by the National Academies National Research Council Board on Science Education. http://dev.tidemarkinstitute.org/sites/default/files/documents/Use%20of%20Evidence%20in%20Changinge%20Undergraduate%20Science%20Education%20%28Austin%29.pdf

Ausubel, D. (1968). *Educational psychology: A cognitive view.* New York, NY: Rinehart & Winston.

Bensimon, E. M., Ward, K., & Sanders, R. (2000). *The department chair's role in developing new faculty into teacher and scholars.* Bolton, MA: Anker.

Bloom, B. S. (1984). *Taxonomy of educational objectives. Handbook 1: Cognitive domain.* New York, NY: Longman.

Bruner, J. (1960). *The process of education.* Cambridge, MA: Harvard University Press.

Chu, D. (2006). *The department chair primer: Leading and managing academic departments.* Bolton, MA: Anker.

Cochran, K. F., DeRuiter, J. A., & King, R. A. (1993). Pedagogical content knowledge: An integrative model for teacher preparation. *Journal of Teacher Education, 44*, 263–272.

Dewey, J. (1897). My pedagogical creed. *School Journal, 54*, 77–80.

Dori, Y. J., & Belcher, J. (2005). How does technology-enabled active learning affect undergraduate students' understanding of electromagnetism concepts? *The Journal of the Learning Sciences, 14*(2), 243–279.

Eraut, M. (1994). *Developing professional knowledge and competence.* London, UK: Falmer Press.

Fairweather, J. (1996). *Faculty work and public trust: Restoring the value of teaching and public service in American academic life.* Boston, MA: Allyn & Bacon.

Fairweather, J. (2008). Linking evidence and promising practices in science, technology, engineering, and mathematics (STEM) undergraduate education: A status report for the National Academies National Research Council Board on Science Education. Commissioned paper for the national academies workshop: Evidence on promising practices in undergraduate Science, Technology, Engineering, and Mathematics (STEM) education.

Fernandez-Balboa, J. M., & Stiehl, J. (1995). The generic nature of pedagogical content knowledge among college professors. *Teaching and Teacher Education, 11*(3), 293–306.

Freeman, S., Eddy, S. L., McDonough, M., Smith, M. K., Okoroafor, N., Jordt, H., & Wenderoth, M. P. (2014). Active learning increases student performance in science, engineering, and mathematics. *Proceedings of National Academic Science U S A.* doi: 10.1073/pnas.1319030111

Freeman, S., Haak, D., & Wenderoth, M. P. (2011). Increased course structure improves performance in introductory biology. *CBE Life Sciences Education, 10*, 175–186.

Freeman, S., O'Connor, E., Parks, J. W., Cunningham, M., Hurley, D., Haak, D., . . . Wenderoth, M. P. (2007). Prescribed active learning increases performance in introductory biology. *CBE Life Sciences Education, 6*, 132–139.

Gates, S. J., & Mirkin, C. (2012). Engage to excel. *Science, 335*(6076), 1545.

Geddis, A. N., Onslow, B., Beynon, C., & Oesch, J. (1993). Transforming content knowledge: Learning to teach about isotopes. *Science Education, 77*(6), 575–591.

Gess-Newsome, J. (1999). Teachers' knowledge and beliefs about subject matter and its impact on instruction. In J. Gess-Newsome & N. G. Lederman (Eds.), *Examining pedagogical content knowledge: The construct and its implications for science education*. Dordrecht, The Netherlands: Kluwer.

Golde, C. M., & Dore, T. M. (2001). *At cross purposes: What the experiences of doctoral students reveal about doctoral education*. Philadelphia, PA: Pew Charitable Trusts. Retrieved from www.phd-survey.org

Grossman, P. L. (1990). *The making of a teacher: Teacher knowledge and teacher education*. New York, NY: Teachers College Press.

Handelsman, J., Ebert-May, D., Beichner, R., Bruns, P., Chang, A., DeHaan, R., . . . Wood, W. B. (2004). Scientific teaching. *Science, 304*(5670), 521–522.

Handelsman, J., Miller, S., & Pfund, C. (2007). *Scientific teaching*: W.H. Freeman & Company in collaboration with Roberts & Company Publishers.

Hashweh, M. Z. (2005). Teacher pedagogical constructions: A reconfiguration of pedagogical content knowledge. *Teachers and Teaching: Theory and Practice, 11*(3), 273–292.

Henderson, C., Beach, A., & Finkelstein, N. (2011). Facilitating change in undergraduate STEM instructional practices: An analytic review of the literature. *Journal of Research in Science Teaching, 48*(8), 952–984.

Henderson, C., & Dancy, M. H. (2008). Physics faculty and educational researchers: Divergent expectations as barriers to the diffusion of innovations. *American Journal of Physics, 76*(1), 70–91.

Injaian, L., Smith, A. C., German Shipley, J., Marbach-Ad, G., & Fredericksen, B. (2011). Antiviral drug research proposal activity. *Journal of Microbiology & Biology Education, 12*, 18–28.

Jensen, J. L., & Lawson, A. (2011). Effects of collaborative group composition and inquiry instruction on reasoning gains and achievement in undergraduate biology. *CBE Life Sciences Education, 10*, 64–73.

Kitchen, E., Bell, J. D., Reeve, S., Sudweeks, R. R., & Bradshaw, W. S. (2003). Teaching cell biology in the large-enrollment classroom: Methods to promote analytical thinking and assessment of their effectiveness. *CBE Life Sciences Education, 2*, 180–194.

Knight, J. K., & Wood, W. B. (2005). Teaching more by lecturing less. *CBE Life Sciences Education, 4*, 298–310.

Leaming, D. R. (1998). *Academic leadership: A practical guide to chairing the academic department*. Bolton, MA: Anker.

Lederman, N. G., & Gess-Newsome, J. (1999). Reconceptualizing secondary science teacher education. In J. Gess-Newsome & N. G. Lederman (Eds.), *Examining pedagogical content knowledge: The construct and its implications for science education*. Dordrecht, The Netherlands: Kluwer.

Loughran, J., Berry, A., & Mulhall, P. (2006). *Understanding and developing science teachers' pedagogical content knowledge*. Rotterdam, The Netherlands: Sense.

Magnusson, S., Krajcik, L., & Borko, H. (1999). Nature, sources and development of pedagogical content knowledge. In *Examining pedagogical content knowledge* (pp. 95–132). Dordrecht, The Netherlands: Kluwer.

Massy, W., Wilger, A., & Colbeck, C. (1994). Department cultures and teaching quality: Overcoming "hallowed" collegiality. *Change, 26*, 11–20.

National Research Council (NRC). (2012). *Discipline-based education research: Understanding and improving learning in undergraduate science and engineering*. Washington, DC: National Academies.

Park, S., & Oliver, S. (2008). Revisiting the conceptualisation of pedagogical content knowledge (PCK): PCK as a conceptual tool to understand teachers as professionals. *Research in Science Education, 38*, 261–284.

Piaget, J. (1954). *The construction of reality in the child*. New York, NY: Routledge.

President's Council of Advisors on Science and Technology (PCAST). (2012). Engage to excel: Producing one million additional college graduates with degrees in science, technology, engineering, and mathematics. Available at www.whitehouse.gov/sites/default/files/microsites/ostp/pcast-engage-to-excel-final_2-25-12.pdf

Senkevitch, E., Marbach-Ad, G., Smith, A. C., & Song, S. (2011). Using primary literature to engage student learning in scientific research and writing. *Journal of Microbiology and Biology Education, 12*, 144–151.

Seymour, E., & Hewitt, N. M. (1997). *Talking about leaving: Why undergraduates leave the sciences*. Boulder, CO: Westview Press.

Shulman, L. S. (1986). Paradigms and research programs in the study of teaching: A contemporary perspective. In *Handbook of research on teaching* (pp. 3–36). York, UK: Macmillan.

Shulman, L. S. (1987). Knowledge and teaching: Foundations of the new reform. *Harvard Educational Review, 57*(1), 1–22.

Smith, D. C., & Neale, D. C. (1989). The construction of subject matter knowledge in primary science teaching. *Teaching and Teacher Education, 5*, 1–20.

Smith, M. K., Wood, W. B., Krauter, K., & Knight, J. K. (2011). Combining peer discussion with instructor explanation increases student learning from in-class concept questions. *CBE Life Sciences Education, 10*, 55–63.

Stufflebeam, D. L., Madaus, G. F., & Kellaghan, T. (2000). *Evaluation models: Viewpoints on educational and human services evaluation* (Vol. 49). Berlin, Germany/New York, NY: Springer.

Tamir, P. (1988). Subject matter and related pedagogical knowledge in teacher education. *Teaching and Teacher Education, 4*, 99–110.

Udovic, D., Morris, D., Dickman, A., Postlethwait, J., & Wetherwax, P. (2002). Workshop biology: Demonstrating the effectiveness of active learning in an introductory biology course. *Bioscience, 52*, 272–281.

van Driel, J. H., Verloop, N., & De Vos, W. (1998). Developing science teachers' pedagogical content knowledge. *Journal of Research in Science Teaching, 35*, 673–695.

Walker, J. D., Cotner, S. H., Baepler, P. M., & Decker, M. D. (2008). A delicate balance: Integrating active learning into a large lecture course. *CBE Life Sciences Education, 7*, 361–367.

Wieman, C. (2007). Why not try a scientific approach to science education? *Change*.http://www.changemag.org/Archives/Back%20Issues/September-October%202007/index.html

Wieman, C., Perkins, K., & Gilbert, S. (2010). Transforming science education at large research universities: A case study in progress. *Change*.http://www.changemag.org/Archives/Back%20Issues/March-April%202010/transforming-science-full.html

Wiggins, G. P., & McTighe, J. (1998). *Understanding by design*. Alexandria, VA: Association for Supervision and Curriculum Development.

Chapter 2
Building Interest and Engagement Through Enrichment Activities

I love going to the talks. I'm very excited about trying things.
I like hearing about the new things other people are doing.

–Biology faculty member

The Teaching and Learning Center (TLC) offers a variety of enrichment programs that bring people together to learn from experts and from one another. This menu of opportunities provides faculty, postdoctoral fellows (postdocs), and graduate students with a venue to learn about effective teaching practices, and fosters dialogue about teaching and learning. Our enrichment programs generally take one of two forms: (1) visits from prominent Visiting Teacher/Scholars who come to share their expertise and engage in conversations, presentations, and brainstorming sessions; and (2) seminars and workshops that are organized by the TLC with university personnel serving as presenters and/or facilitators. In this chapter, we will describe both types of enrichment programs and provide examples of each.

Visiting Teacher/Scholars

Every semester, the TLC hosts a Visiting Teacher/Scholar who is nationally recognized for his or her ability to integrate effective teaching and science research. Our intent is to showcase individuals who demonstrate that excellence in teaching can coexist with excellence in scientific research.

Characteristics of Visiting Teacher/Scholars

Science research, science education research, and higher education research are distinct disciplinary fields that are often isolated (Henderson, Beach, & Finkelstein, 2011). Our Visiting Teacher/Scholars are the exceptional scholars with formal academic training in the sciences whose areas of expertise and research focus span two or more of these generally discrete specializations.

Our Visiting Teacher/Scholars' research focus can be characterized as fitting into one of three broad categories that represent the overlap between science education

© Springer International Publishing Switzerland 2015   19
G. Marbach-Ad et al., *A Discipline-Based Teaching and Learning Center*,
DOI 10.1007/978-3-319-01652-8_2

research, science research, and higher education research (see text boxes below for specific examples). These categories are representative rather than exhaustive, and they reflect patterns of ongoing scholarship as well as the types of visitors who are likely to have a positive impact on promoting teaching and learning in the context of our College.

The first category of Visiting Teacher/Scholars consists of individuals who have active research programs in both science and science education. This research combination is rare, and these Teacher/Scholars can serve as role models or inspiration to research faculty in our College. Most of our faculty members have traditional tenure-track career trajectories, and many are skeptical that it is possible to engage in high quality research in both science and science education. The success of these Teacher/Scholars illustrates the possible synergy of science and science education research. Furthermore, we have found that Visiting Teacher/Scholars with dual research foci attract a broader range of seminar attendees, as they can connect with faculty members from both science and education disciplines. The science research of the Visiting Teacher/Scholar serves as a 'hook' to draw faculty members, postdocs, and graduate students who are interested in the scientific research area to a session on teaching and learning that they might not otherwise attend.

Jo Handelsman, a Howard Hughes Medical Institute Professor, visited us from the Department of Molecular, Cellular and Developmental Biology at Yale University. Handelsman, a microbiologist with more than 100 scientific research publications, is also widely renowned for her work on scientific teaching, professional development for current and future biology faculty members, and science education policy for higher education (Handelsman et al., 2004; Handelsman, Miller, & Pfund, 2007). Her science research drew many of our faculty members to attend her seminar and small meetings. In these interactions, Handelsman shared her vision on postsecondary STEM education and exposed our faculty to national policy conversations related to undergraduate STEM education and reform.

I strongly believe my teaching style has benefited from my interactions with [Dr. Handelsman]. She has given me numerous tips and strategies for the classroom!

–Graduate student in Chemistry

The second category of Teacher/Scholar consists of faculty members who began their academic careers as research scientists but now focus primarily on science education research (Bush et al., 2008). These individuals are usually national leaders in developing and implementing innovative practices in teaching and learning. They generally have published extensively on their teaching and learning initiatives, won awards for these initiatives, and developed model programs that can be widely replicated.

Diane Ebert-May, a biology education researcher who visited us from the Department of Plant Biology at Michigan State University, has published extensively on teaching and learning in undergraduate biology courses. Her science education research interests include the incorporation of evidence-based active learning approaches in large enrollment undergraduate biology courses and professional development for future faculty (Ebert-May et al., 2011; Ebert-May & Hodder, 2008; Wyse, Long, & Ebert-May, 2014). Many of our faculty members are in the process of redesigning their own large enrollment courses, and were interested in learning from Ebert-May about overcoming the many barriers to implementing innovative, learner-centered approaches in this type of course.

It was just fascinating to listen to [Diane Ebert-May's] talk ... hearing her talk about stuff she's done. [These talks] always get me excited about trying to do something new or different.

–Biology instructor

The third category includes higher education administrators who have initiated major STEM reforms and conducted research on or evaluated those initiatives. These administrators generally have a background in scientific research, but have subsequently assumed major campus and/or national leadership roles. They provide a perspective on change initiatives that go beyond departmental and sometimes even college boundaries. Their science background makes their perspective more relevant to our faculty members, postdocs, and graduate students than would be the case with administrators with non-science backgrounds.

At the time of her visit, **Claudia Neuhauser** was the Vice Chancellor for Academic Affairs at the new campus of the University of Minnesota, Rochester. In this role, she developed and evaluated innovative interdisciplinary degree programs and courses, and promoted ongoing education collaborations across disciplines. Neuhauser is a mathematician by training, and her research interests have spanned areas of biology across levels of organization, from the genome to ecological communities. During her visit, Neuhauser met with individuals and groups who are engaged in ongoing initiatives to enhance interdisciplinary connections between biology, chemistry, physics, and mathematics.

The meeting with Claudia was interesting. [We discussed] our intention about scaling up and how we imagined doing that ... [and] our design strategies.

(continued)

She was very positive about what we're doing ... and having her reflection on our decisions about interdisciplinary education will be useful in moving forward.

–Postdoctoral research fellow who works on interdisciplinary initiatives in the University of Maryland (UMD) Physics Education Research Group

In selecting Visiting Teacher/Scholars, we strive for diversity in scientific disciplines to represent the four departments that we serve. We also seek Teacher/Scholars who are engaged in change initiatives that corresponded closely to our own. In this way, the Teacher/Scholars are particularly relevant for our audience, and their visits can help establish extended collaborations.

I have attended nearly every talk in your Visiting Teacher/Scholar series and met with many of the visiting scholars. As I had recently moved from a pure research institute to an academic environment where teaching is central, this series was instrumental to my teaching. From models of teaching and engaging students to creating opportunities for active learning in the classroom, the topics of the series have been incredibly relevant to every aspect of my teaching activities.

–Biology faculty member

Activities with Visiting Teacher/Scholars

Visiting Teacher/Scholars typically spend two days on campus. During that time, they offer a seminar on teaching and learning and meet with faculty, postdocs, and graduate students individually and in small groups. Some Visiting Teacher/Scholars also give a second seminar that focuses on their current scientific research. As much as possible, meals with the Visiting Teacher/Scholars are planned as community events, and generally include dinner with faculty members and lunch with graduate students. Having many meetings of different types provides the opportunity for large numbers of faculty members, postdocs, and graduate students to interact with the visitors.

It always feels relieving to hear from a visiting scholar that they started off just like us, graduate students with a vocation and determination. That gives us hope that someday we can make it to stardom, like they have.

–Graduate student in Chemistry

The visit typically begins with a breakfast meeting with the TLC staff, who provide an introduction to the history of the TLC, the objectives of the Visiting Teacher/Scholar program, and background on science education initiatives currently underway in the College. We have found this to be a helpful orientation for visitors because of the uniqueness of our disciplinary center. This introduction allows them to understand our institutional context and tailor their remarks accordingly. The meeting also provides opportunity for the TLC staff to seek advice from the Visiting Teacher/Scholar on professional development, curriculum development, and institutional transformation quandaries.

For additional information on logistics and preparations for hosting a Teacher/Scholar, see the Tip Sheet in the Implementation Guide. For a historical list of all TLC Visiting Teacher/Scholars, see cmns-tlc.umd.edu/VTS.

Seminars and Workshops

The second type of enrichment activity consists of seminars and workshops hosted by the TLC with university personnel serving as presenters and/or facilitators. These seminars and workshops provide a low-distraction environment to introduce participants to innovative practices and develop their expertise in science teaching. The duration of a given program varies depending on content and objectives, with seminars typically lasting about an hour and a half and workshops generally lasting between one and four hours.

Given the many responsibilities that our audience must balance, the short format of the seminars and workshops makes them accessible and enables us to introduce these topics to relatively large groups. Additionally, our enrichment activities sometimes serve as gateway activities in that they introduce faculty members to the TLC's activities and to topics in teaching and learning. Conversations and interests that are initiated in seminars and workshops often continue in different settings. These interests may be pursued through more in-depth professional development, in long-term course redesign collaborations between faculty members who teach interconnected courses, and in other reform initiatives.

In this book, we refer to **seminars** as professional development activities in which one or more experts brings new knowledge to a group with a shared interest in a topic, with most of the meeting time dedicated to the experts' presentation of this topic. While the experts' sharing of knowledge is the primary focus of the seminar, it may also include an opportunity for audience discussion or question and answer sessions following the presentation.

Workshops differ from seminars in that the presentation component is shorter and serves to provide a basic introduction to a topic. Workshop participants actively engage in one or more activities related to the topic during the majority of the meeting time.

Guiding Principals in Developing Seminars and Workshops

Our seminars and workshops are intended to follow best practices in adult learning. Many aspects of our workshops reflect Knowles' (1980) principles of adult learning:

- *Adult learners are internally motivated and self-directed.* Participants play an active role in our seminars and workshops, and the format is designed to facilitate their learning.
- *Adult learners have accumulated a wealth of knowledge through their experiences.* Our workshops build upon participants' existing knowledge of teaching and learning, and frequently draw upon participants' teaching experiences.
- *Adult learners are goal-oriented and practical.* Our programs have clearly defined goals related to our participants' tasks and challenges as instructors.
- *Adult learners are relevancy-oriented.* Through our disciplinary focus, we ensure that our programming is highly relevant to the work of our audience.
- *Adult learners should be respected.* We provide detailed information on the objectives and content of the upcoming seminars and workshops (e.g., detailed abstracts, readings, background information on guest speakers, and links to online resources) so that potential attendees know exactly what they can gain from attending.

The Value of Offering Food
We have found it very important to provide food at our workshops and seminars. Providing food serves multiple purposes: the food is an incentive to attend; it creates a collegial, community atmosphere; and it demonstrates institutional support.

Selecting Seminar and Workshop Topics

One of the most important aspects of planning seminars and workshops is determining the topics to be covered, and ensuring that these topics will appeal to our audience and be relevant to their work and professional development needs. In planning specific topics for our workshops and seminars, we consult with our stakeholders (i.e., dean, department chairs, faculty members, graduate students) to learn what topics interest them. TLC staff also recommend topics that seem relevant and interesting to our stakeholders, such as current College curricular revisions.

Evaluation of Seminars and Workshops

Following some seminars and workshops, we request that participants complete an evaluation survey. Such surveys generally include items with a rating scale as well as open-ended questions. Through these evaluations, participants have provided helpful feedback that has enabled the TLC to better tailor seminar and workshop topics, formats, and materials to the needs and preferences of our audience.

PCK Focus of Seminars and Workshops

All TLC enrichment programs are focused on teaching and learning specifically as it applies to biology and chemistry. Due to this disciplinary focus, our programs provide pedagogical content knowledge (PCK) rather than general pedagogical knowledge. We offer a broad range of topics within this targeted content area, and our workshops span the five components of PCK: (1) student understanding of science, (2) science curriculum, (3) instructional strategies, (4) assessment of student learning, and (5) orientation to teaching science. See Chap. 1 for more details about PCK and its components.

> *I'm glad that we have a disciplinary teaching and learning center. I did go to a lot of workshops offered by the campus Center for Teaching Excellence, but it was so skewed to humanities and not sciences, and labs are [a] totally different beast from discussion sections.*
>
> –Biology instructor

In the subsequent sections, we describe each PCK component, and provide an example of a seminar or workshop that addresses the component. The Implementation Guide provides detailed descriptions of these sample workshop and seminars, and includes seminar/workshop timelines, activities, materials, and suggestions for supplemental readings.

PCK 1: Student Understanding of Science

The first PCK component refers to knowledge about how to assess and address students' prior knowledge, construction of knowledge, and diversity. The theory of constructivism suggests that learning occurs as students iteratively connect new knowledge to prior knowledge, which is built through previous academic work as well as experiences in everyday life (Ausubel, 1968; Bruner, 1960; Mintzes, Wandersee, & Novak, 2005; Piaget, 1954; Vygotsky, 1978). This prior knowledge may be valid in that it aligns with widely accepted scientific theories, or it may be inconsistent with accepted theories, in what is commonly termed an alternative conception (Fisher, 1983; Gilbert, Osborne, & Fensham, 1982; Thijs & van den Berg, 1993). Empirical studies indicate that prior knowledge strongly impacts academic success (Bloom, 1976; Dochy, Segers, & Buehl, 1999; Marzano, 2004; Tobias, 1994). Therefore, it is important for educators to be aware of and adjust their teaching based on students' prior knowledge.

> **Alternative conceptions** or 'misconceptions,' are commonly held ideas that are inconsistent with a concept as it is understood by scientists (Thijs & van den Berg, 1993). Alternative conceptions have a variety of generative causes (Fisher, 1983; Gilbert et al., 1982; Marbach-Ad, 2009; Thijs & van den Berg, 1993), such as arising from everyday life experiences or students' overgeneralization. It is important to understand the generative causes of alternative conceptions, because this helps to target learning strategies to overcome them (Krause, Kelly, Tasooji, Corkins, & Purzer, 2010).

In any given class, students vary in terms of their prior knowledge and framework for integrating new knowledge. This diversity may stem from different cultural backgrounds, academic trajectories, retention of prior learning, and learning styles (Birenbaum & Dochy, 1996; Magnusson, Krajcik, & Borko, 1999). However, faculty members do not always address this diversity. In some cases, faculty members teach as if all students enter their course with no relevant prior knowledge. In other cases, faculty members work under the assumption that all students bring sufficient knowledge from prior coursework. Neither assumption is likely to be true, so it is valuable for faculty members to verify their assumptions about student prior knowledge.

In the departments that we serve, we find it particularly important to probe students' prior knowledge in first- and second-year introductory courses. For second-year courses this is especially important because students come via multiple paths; some have earned Advanced Placement credit in high school that allows them to move directly into second-year courses, others may have transferred from two- or four-year institutions, and still others have taken prior coursework in our own institution. Each of these paths has implications for what students were previously taught. Instructors should also probe prior knowledge in upper level courses, due to differences in students' prior coursework and content retention.

In our professional development programming related to this PCK component, we have two overarching goals. First, we seek to raise awareness of the importance of understanding and addressing students' prior knowledge. Second, we aim to expose participants to varied tools that they can use to gauge their students' prior knowledge and to support them in using these tools.

In seminars and workshops related to this PCK component, presenters or facilitators share strategies for gauging students' prior knowledge, such as pretests, surveys, interviews, and classroom activities. For example, one faculty member presented an innovative mechanism for gauging students' prior knowledge through concept mapping (described more fully in the Implementation Guide). This workshop provided an overview of common alternative conceptions in the chemical and life sciences, and offered strategies for overcoming those alternative conceptions. We began with an introduction and definition of key terms related to alternative conceptions. Participants then engaged in small group activities to identify students' alternative conceptions, investigate their causes, and explore instructional techniques to overcome them. The workshop concluded with several brief research presentations on identifying and addressing prevalent alternative conceptions held by high school chemistry and undergraduate biology students.

PCK 2. Science Curriculum

The second PCK component refers to knowledge of curriculum within and across courses. Curriculum is a term with a nebulous meaning and many definitions (Brandwein, 1977; Egan, 2003; Wiggins & McTighe, 1998). Wiggins and McTighe (1998) define curriculum as "the specific blueprint for learning that is derived from *desired results*—that is, content and performance standards" that may be locally or externally determined (pp. 5–6; emphasis in original). Literature on curriculum emphasizes that it is not just a list of topics to be covered, but an organizing framework for content delivery and a map for how and when to teach desired content (Graff, 2011; Kauffman, Moore, Kardos, Liu, & Peske, 2002; Wiggins & McTighe, 1998). Here, we refer to curriculum as the topics that are taught in a single course or across courses. Our conceptualization of curriculum encompasses how topics are taught only in a very general way, and does not include the level of specificity of lesson plans, instructional strategies, and instructional materials, which we include in PCK 3.

Developing curriculum should be a deliberate and systematic process. Wiggins and McTighe (1998) recommend using a backward design approach to building curriculum. This design involves first identifying desired results or learning goals, then determining acceptable evidence for whether students have achieved the desired results, and only then sequencing topics and planning instructional activities (Wiggins & McTighe, 1998). This backward design approach, when applied to individual courses and sequences of courses, promotes coherence in what students learn and how they assimilate into a discipline.

Curricular knowledge is generally understood to have two dimensions: vertical and horizontal (Shulman, 1986). Vertical curriculum refers to what is taught in a subject area across courses and years of study. Horizontal (or lateral) curriculum refers to what is taught simultaneously, whether in parallel courses (e.g., multiple sections of introductory biology) or in related courses that students take in any given semester (e.g., introductory biology and chemistry courses). Course instructors should be aware of what is commonly taught in related courses, whether vertically or horizontally aligned with their own course.

As a professional development providers in a large university, we work with faculty members who teach related courses as well as faculty members who teach multiple sections of a single course to align learning goals and materials across the spectrum of undergraduate study. In both cases, much of our professional development focuses on student learning progressions, and how curriculum alignment can promote coherence and the development of knowledge. To respect faculty members' academic independence, we collaborate with them to build an understanding of best practices in curriculum development and support them as they apply these practices in their own courses. In accordance with national recommendations (AAAS, 2011), our approach emphasizes depth over breadth of content coverage. We also promote the use of empirically validated, content-specific learning progressions to guide curriculum development.

Learning progressions are "research-based cognitive models of how learning of scientific concepts and practices unfolds over time" (Duncan & Rivet, 2013, p. 396). They map pathways that students may follow in developing more sophisticated understanding of key ideas and skills necessary for science literacy (Fortus & Krajcik, 2012). Using learning progressions to guide curriculum design supports coherence within and across courses (Duschl, Schweingruber, & Shouse, 2007).

In seminars and workshops related to this component of PCK, we address topics such as student learning progressions, content and course sequencing, and strategies for emphasizing conceptual understanding rather than content coverage. These seminars and workshops sometimes address curricular changes within a single course, and may include brief presentations from faculty members on how and why they changed the topics and/or sequencing of topics within their course. Other seminars and workshops address coordinated changes across multiple interlinked courses. In all of these programs, we emphasize how faculty members can collaborate to better understand the knowledge that students bring from one course to another, so that faculty members can more effectively bridge student understanding across courses.

The Implementation Guide describes one of our workshops on initiating, coordinating, and implementing curricular innovations. In this workshop, we provided two examples of recent large-scale curricular changes from our College: reordering introductory general and organic chemistry classes to move organic chemistry earlier in the sequence and building a curricular continuum between nine undergraduate microbiology courses. As part of the workshop, we encouraged more changes of this type and provided examples of how the TLC can support these changes.

PCK 3. Instructional Strategies

The third component of PCK refers to an awareness of instructional strategies and an understanding of how to employ those strategies appropriately and effectively. In recent years, emerging research and national policy recommendations have strongly supported the use of student-centered instructional approaches (AAAS, 2011; Freeman et al., 2007, 2014; Marbach-Ad & Sokolove, 2000; NRC, 2003; Smith, Wood, Krauter, & Knight, 2011). These approaches reflect a perspective on the purpose of higher education as not focused on transmitting knowledge, but rather on creating an environment in which learners are engaged in their learning and construct their own knowledge (Handelsman et al., 2007; Wieman, 2007).

Research focused on teaching in the sciences highlights the importance of teaching science with the same rigor that is applied to scientific research. This idea, termed scientific teaching, suggests that science educators should employ evidence-based teaching approaches and reiteratively assess the effectiveness of those approaches (Handelsman et al., 2004, 2007). Instructors at the university level are well equipped to employ the scientific teaching approach because of their extensive training as researchers.

Seminars and workshops centered on this component of PCK focus on expanding the use of evidence-based, student-centered instructional strategies, including inquiry-based learning, cooperative and collaborative learning, and technology-aided instruction. In some cases, these workshops and seminars consist of presentations by science instructors in which they provide specific examples from their own classroom. In other cases, the seminars and workshops are led by science education specialists and focus on research about the effectiveness of and appropriate uses for different evidence-based strategies.

One of the TLC's recent seminars in this theme dealt with blended learning and different ways in which blended learning techniques can be used productively in science classes. Administrators as well as faculty members have expressed interest in increasing blended learning offerings in our undergraduate program. This seminar complemented professional development offered by the campus Center for Teaching Excellence and Office of Learning Technologies. In our seminar, four faculty members served as panelists and shared their experiences using different configurations of blended learning in different types of courses. After this panel presentation, the entire group engaged in a conversation about the possibilities for and implications of incorporating blended learning in science coursework. We describe this seminar in more detail in the Implementation Guide.

Blended learning combines face-to-face instruction with computer-mediated or online learning. The online approach may be synchronous (live) or asynchronous (self-paced), and often includes a combination of the two. Blended learning can use a variety of media and incorporate different degrees of interactivity. It may principally include expository instruction via recorded lectures and static materials, or it may foster a more constructivist approach through activities such as scenario-based learning experiences and simulated modeling. In the online component of blended learning, students often interact with each other and with the instructor, which offers an opportunity to increase the number and frequency of those interactions without increasing face-to-face instruction. The online component may include quizzes or other assessments, differential supports based on student performance, and some degree of learner control over which instructional activities and how much instruction are accessed.

PCK 4. Assessment of Student Learning

The fourth component of PCK addresses methods and instruments for assessing student learning. As the backward design approach suggests, assessments should be aligned with learning goals and instructional activities (Handelsman et al., 2007; Wiggins & McTighe, 1998). Classroom assessment helps instructors to better understand how well their students are learning course materials, and allows them to adjust their teaching to help students progress in the material and become better learners (Angelo & Cross, 1993). Assessment also helps students to understand their progress in learning and to adjust their learning efforts and strategies appropriately (Handelsman et al., 2007). Different methods and instruments can play a complementary role in aligning to different learning goals and providing different types of feedback.

The large body of literature on assessment uses multiple typologies. Some literature separates assessment into categories based on its purpose. Angelo & Cross (1993), for example, describe three broad purposes: assessing course-related knowledge and skills; assessing learner attitudes, values, and self-awareness; and assessing learner reactions to assessment. Another common categorization of assessment highlights the differences between formative and summative assessments (Black & Wiliam, 1998; Middle States Commission on Higher Education, 2007; Sadler, 1989). Formative assessments measure students' learning progress throughout a course or course of study and are usually used internally. The goal of formative assessments is to provide feedback to instructors and to students to improve the performance of individual students rather than to judge their performance. Summative assessments generally occur at the end of a teachable unit, course, or program, and may be used for internal, external, or accountability purposes. The primary goal of summative assessment is to judge student performance and to determine whether students achieved overall goals associated with the unit, course, or program. Formative and summative assessments play complementary roles in improving student learning and the effectiveness of instruction.

Different assessment instruments and item types can be used depending on the purpose of the assessment, since different types may be more appropriate for some learning goals than for others (Magnusson et al., 1999). Professional development in this area can introduce instructors to the wide variety of available instruments and items. Assessment tools should go through a validation process to ensure their validity and reliability. While some assessments may be ungraded, most assessments are graded. Grading techniques vary based on the assessment and item type. Multiple-choice items tend to be relatively straightforward and efficient to grade, but are not appropriate in every context. Open-ended items such as constructed response items, portfolios, and presentations are more difficult to grade. Grading rubrics promote objectivity in grading open-ended items and can increase inter-grader reliability.

Databases such as the Field-tested Learning Assessment Guide (FLAG) serve as valuable resources for instructors to use to explore different assessment types and access validated assessment tools and grading rubrics (www.flaguide.org). Instructors within a department may also collaborate with their peers to validate tools that they develop and/or share assessments or items across sections or even courses.

Assessment is intertwined with other PCK components and therefore is addressed indirectly in a number of our seminars and workshops. In this section, we refer only to seminars and workshops with a primary focus on assessing student learning. In these seminars and workshops, we share assessment tools; explore assessments that measure different levels of student understanding; and discuss different mechanisms for grading, such as using rubrics to evaluate open-ended questions. These seminars and workshops provide a venue for introducing faculty to a wider range of assessment practices than the traditional exams that most faculty members rely on, and give them concrete ideas for how differing assessments can be implemented effectively and efficiently.

In the Implementation Guide, we describe a workshop led by a nationally recognized expert in assessing undergraduate students' understanding of science. This workshop provided a broad overview of the types and purposes of assessment, as well as a hands-on activity to explore different types of assessment tools that may be used to uncover how students understand complex course content. Assessment knowledge is considered to be the highest stage in teacher professional growth (Avargil, Herscovitz, & Dori, 2012). Given the complexity of knowledge of assessment, we brought in an outside expert to lead this workshop and introduce our faculty to current research and theory in this area.

PCK 5. Orientation to Teaching Science

This PCK component addresses knowledge and beliefs about the way science should be taught and the goals of teaching science. The theory of scientific teaching (Handelsman et al., 2004, 2007) has increasingly shaped orientations to teaching science around its call to approach the teaching of science with the rigor and reliance on evidence that characterize scientific research. The graduate training that future faculty members receive builds their identity as scientific researchers who approach their scientific research with emphasis on rigor and empirical evidence. This graduate training generally does not equally focus on building future faculty members' identities as science teachers. This identity as a science teacher shapes much of the instruction that occurs in classroom (Kember & Kwan, 2000; Magnusson et al., 1999), but may be underdeveloped in new instructors.

Our professional development in this area is designed to support faculty members as they shape their identities as reflective, reform-minded teachers. Rather than suggesting what this identity should be, we provide opportunities for faculty members to explore what they think a reflective, reform-minded teacher looks like and to foster this identity in themselves. Developing this identity prepares teachers to face the unique challenges that their profession entails, particularly in light of the national call to improve undergraduate science education.

In seminars and workshops that address this area of PCK, we cover topics including how to identify what constitutes effective teaching, the expected outcomes of effective teaching, and how to approach teaching in a way that promotes optimal outcomes. Examples of such seminars and workshop topics include developing a teaching philosophy, discussing prominent reform recommendations for improving science education, and identifying desired outcomes for teaching and learning. Many of these seminars and workshops provide an opportunity for participants to engage in reflection about their own teaching as well as to consider how best practices and national reform efforts can and should be incorporated in their teaching.

In the Implementation Guide, we share an example of a workshop in which biology faculty members collaborated in the identification of the hallmarks of effective teaching, which informed the creation of a rubric for peer review of teaching. This three-and-a-half hour workshop was held during a faculty retreat for the biology department. The purpose of the workshop was to engage the entire department in a discussion of what constitutes effective teaching and how effective teaching is manifested in the classroom. The workshop was part of an extended process to create a systematic approach to peer review within the department, led by the department chair with assistance from a small committee of faculty.

Conclusion

The activities described in this chapter served as gateway activities through which many faculty members, postdocs, and graduate students were introduced to the services of the TLC and to topics related to teaching and learning. Given this gateway role, it was particularly important that the focus of Visiting Teacher/Scholars, seminars, and workshops reflected the target audience's needs and interests. The examples we mention above and describe in the Implementation Guide were specifically designed for our audience and can be adapted as appropriate for other contexts.

Implementation Guide

The Implementation Guide includes tips for Teacher/Scholar visits, as well as outlines of sample seminars and workshops. These materials may be used and adapted as needed.

Tips for Hosting a Successful Teacher/Scholar Visit

Selecting and Recruiting Visiting Teacher/Scholars

- Seek suggestions from college and departmental leadership, as well as faculty members who are active in change initiatives.
- Seek suggestions from teaching and learning specialists in the university, e.g., director of campus teaching and learning center.
- Draw upon your professional networks to get recommendations for emerging Teacher/Scholars.
- Take advantage of related events in your area and invite Teacher/Scholars who may be visiting your area for other reasons. This may provide an opportunity to save on travel expenses and to bring Teacher/Scholars who may not otherwise have availability to plan an independent visit.
- Plan early and be flexible. In our experience, many prominent Teacher/Scholars need ample advance notice—at least a semester and sometimes as much as a year—to schedule their visit. If the suggested semester does not work for the visit, a future semester may.
- Give as much detail about the visit as possible in the invitation letter. The letter should specify length of visit, possible timeframes, audience, research areas of interest, and funding support if available.

Publicizing the Visit

– After scheduling a visit, request a seminar title, abstract, and short biosketch from the Teacher/Scholar. This information is needed to begin publicizing the visit.
– Publicize the visit through multiple avenues and to multiple audiences. We generally create a webpage on the TLC website (e.g., cmns-tlc.umd.edu/handelsman); send announcements to departmental electronic mailing lists for faculty members, postdocs, and graduate students; and post notices on the campus-wide event calendar.
– Invite faculty members, postdocs, and graduate students to schedule individual or small group meetings with the visitor.
– Send targeted emails to college and university administrators who may be particularly interested in the topic of the seminar and/or meeting with the visitor.
– Seek out specific faculty members or groups of faculty to meet with the visitors, based on your knowledge of their research interests and teaching initiatives.
– Send save-the-date notices a month before the visit and a reminder notice a few days before the visit.

Creating an Agenda of Activities

– Begin the visit with a meeting with teaching and learning center staff, who provide an introduction to the center, a summary of the objectives of the Visiting Teacher/Scholar program, and an overview of relevant science education initiatives currently underway in the college.
– Make the seminar the centerpiece of the visit. Schedule the seminar at a time when most faculty members are not teaching class. We have found 3:00 to 4:30 pm to be the best time, but this may vary based on the institution's class schedule. Book the seminar room as soon as a date for the visit has been selected.
– Reserve at least one meal for graduate students and postdocs to eat with the visitor. Some visitors prepare special remarks for this audience, while others prefer a more casual, open-ended discussion.
– Send the agenda to the visitor at least a few days prior to the visit. Include details about meeting participants and, if a meeting is based around a specific reform or change (e.g., redesign of the introductory chemistry sequence), include relevant information or website links.

Example Seminars and Workshops by PCK Component

On the pages that follow, we share details about the workshops and seminars that were mentioned earlier in the chapter. For each seminar or workshop, we provide an overview of the session structure, content, materials, and interactive activities. These details are intended to illustrate how seminars and workshops can build participants' knowledge of each PCK component and to assist faculty development providers in the implementation of these or similar workshops in their own institutions.

PCK 1: Overcoming Students' Alternative Conceptions in the Chemical and Life Sciences

The TLC Director facilitated this 90-min workshop and invited select course instructors to share their experience and research addressing students' alternative conceptions. In addition to these presentations, participants engaged in small group activities. Workshop components are outlined below and described in the pages that follow.

1. **Introduction and Definitions of Key Terms (5 min)**
2. **Generative Causes of Alternative Conceptions (25 min)**
3. **Identifying and Overcoming Alternative Conceptions (20 min)**
4. **Course Instructor Experience and Research (30 min)**
5. **Question and Answer (10 min)**

1. **Introduction and Definitions of Key Terms**

At the start of the workshop, the facilitator provided definitions of concept, conceptions, and alternative conceptions. The facilitator also discussed key characteristics of alternative conceptions:

- Alternative conceptions are widely shared and may exist across many countries, within a variety of cultural and environmental contexts, although some alternative conceptions may be culturally dependent.
- Alternative conceptions can be similar to views that were held by scientists in the past.
- Alternative conceptions are very persistent (Krause et al., 2010).

Concept is the scientific idea underlying a class of things or events, as currently accepted by the scientific community and documented in leading textbooks. A concept acquires its meaning through its network of relationships with other concepts.

Conception refers to an individual's idea of the meaning of a concept. Such an interpretation would usually have some idiosyncratic features, even if the individual is a scientist.

Alternative Conception, or 'misconception,' refers to a conception that in some aspects is contradictory to, or inconsistent with, the concept as intended by the scientific community. Such inconsistency usually shows in one or more relations of the conception with other conceptions. It thus often involves more

(continued)

than one concept. We only talk of alternative conceptions if alternative ideas have some robustness and persistence across ages and levels of schooling.

–Thijs & van den Berg, 1993

2. Generative Causes of Alternative Conceptions

Participants worked in small groups or individually to complete a worksheet with questions that typically elicit students' alternative conceptions. The questions can be seen in the box; the alternative conceptions that these questions should uncover, as well as explanations of their possible generative causes, can be found in the Annotated Alternative Conceptions Worksheet in Chapter 5.

Alternative Conceptions Worksheet

1. Why do seasons happen? Why is it "hot" in the summer and "cold" in the winter? (Schneps & Sadler, 2003)
2. Why do we have different phases of the moon? (Schneps & Sadler, 2003)
3. Why, in hospitals, might nurses take the plants from patient rooms at night?
4. The balloon was left out in the sun. Why did the balloon pop?
5. Consider a copper wire. Divide it into two equal parts. Divide one half into two equal parts. Continue dividing in the same way. Will this process come to an end? (Stavy & Tirosh, 2000)
6. Two roommates fall ill: one has an ear infection and one has pneumonia. Is it possible that the same causative agent is responsible for both types of disease?

    (a) Yes, because both individuals live in the same room and therefore the source of the infection has to be the same.
    (b) Yes, because the same bacteria can adapt to different surroundings.
    (c) No, because each bacterium would cause one specific disease.
    (d) No, because one infection is in the lung while the other is in the ear.
    (e) I do not know the answer to this question. (Marbach-Ad et al., 2010)

7. Two sugar cubes are added to a bowl containing 20 oz of water. One sugar cube is added to a bowl containing 10 oz of water. Which bowl of water is sweeter?
8. Cats usually have five digits on each paw. Omer has a cat with six digits. Omer's cat gave birth to two kittens that also had six digits. How can you explain this? Was something transmitted from the mother to the kittens? If so, what is it? (Marbach-Ad & Stavy, 2000)

Table 2.1 Generative causes of alternative conceptions and related questions from alternative conceptions worksheet

Generative causes of alternative conceptions	Question(s)
Everyday language/experience	1, 2, 3
Intuition	4, 7
Over-generalization	2, 5, 6
Under-generalization or over-simplification	3
Confusion between extensive and intensive entities	7
Connections between the macro- and micro-level	8

After finishing the worksheet, the entire group reconvened to identify the alternative conceptions that the worksheet may elicit and categorize them based on their possible causes or origins. To guide this discussion, the facilitator provided a list of generative causes of alternative conceptions (Table 2.1).

Following this discussion, the facilitator showed the opening scene from the video "A Private Universe" (Schneps & Sadler, 2003). This scene summarizes characteristics and possible generative causes for students' alternative conceptions in astronomy, as documented through interviews with Harvard graduates, their professors, and a bright ninth-grader who has some confused ideas about the orbits of the planets.

Participants were then divided into homogeneous groups based on their discipline (Biology, Chemistry, Physics, and Mathematics). Each group received one generative cause for alternative conceptions. The groups were tasked with finding examples of alternative conceptions that can emerge from this generative cause, based on their experiences in the classroom. Finally, each group summarized their examples on posters, which were hung up on the walls of the room for other participants to view.

3. **Identifying and Overcoming Alternative Conceptions**

The facilitator discussed potential ways to identify alternative conceptions through different types of assessment:

- Responding to open-ended questions
- Paraphrasing key terms
- Drawing visual representations of concepts and relationships between concepts (e.g., concept mapping)
- Creating word association

After providing this global overview of assessment tools, the facilitator then focused on concept maps. She discussed how concept maps can be used not just to identify alternative conceptions but also to overcome them.

Participants split into groups of about five people to build their own concept map. Each group was given the same set of concepts: living things, plants, animals, molecules, water, motion, states, liquid, solid, gas, and heat (concepts taken from Novak and Gowin (1984), p. 18), with each concept written on a card. The groups then arranged the cards on a large piece of paper in a logical order that allowed them to make as many connections as possible between concepts. They then drew lines between concepts and came up with propositions to describe the relationships between the concepts. Once they were finished, each group hung its concept map on the wall to share with the others.

The whole group then discussed the many possible maps that could be drawn to connect these 11 concepts, and highlighted differences in the concepts maps that the groups had created. The facilitator also shared the two examples below (Novak & Gowin, 1984) to show the different ways that students can conceptualize the same set of concepts. For example, concepts can be arranged from the micro level (molecules) to the macro level (multicellular organisms), as in Fig. 2.1, or from macro level to micro level, as in Fig. 2.2.

4. Faculty experience and research

A panel of three faculty members and a graduate student shared examples of how they have identified alternative conceptions and used concept maps in their classes.

The first presenter was a chemistry educator who shared his research involving high school chemistry students. He found that students had difficulties connecting macro-level phenomena with micro-level explanations, and sometimes incorrectly applied macroscopic ideas to the molecular world. The presenter pointed out that alternative conceptions resulting from these difficulties persist throughout high school and even to the undergraduate level (Stieff, 2005).

The next panelists, the instructor and GTA of an upper-level undergraduate immunology course, presented their research on the use of concept mapping as a voluntary homework exercise in their class. The research team, which consisted

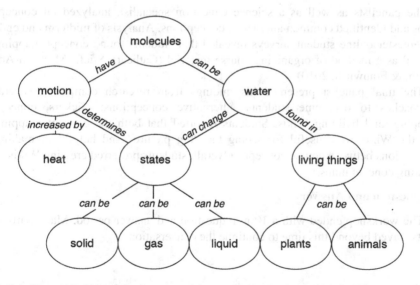

Fig. 2.1 Hierarchical concept map arranged from micro level to macro level. (Novak & Gowin, 1984)

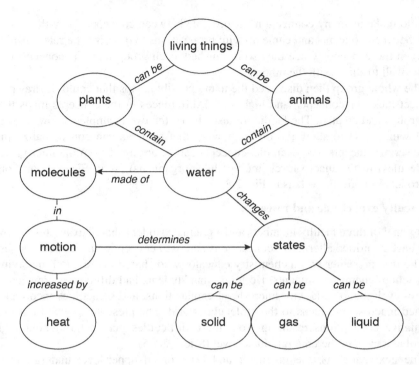

Fig. 2.2 Hierarchical concept map arranged from macro level to micro level. (Novak & Gowin, 1984)

of the panelists as well as a science education specialist, analyzed all concept maps and identified common alternative conceptions. Analysis of midterm and end-of-semester online student surveys revealed that students found concept mapping helpful as a method of organizing course material (Cathcart, Stieff, Marbach-Ad, Smith, & Frauwirth, 2010).

The final panelist presented his findings from research comparing several approaches to overcome students' alternative conceptions: clickers, concept mapping, and building Wikis. Students reported that both the concept mapping and the Wikis were useful for seeing the "big picture" and helpful in making connections between course concepts. Overall, students preferred creating Wikis to creating concept maps.

5. Question and Answer

The workshop ended with a 10-min question and answer period. Many participants stayed beyond this time to continue the conversation.

PCK 2: Undergraduate Curriculum Innovations in the College of Chemical and Life Sciences

This workshop began with faculty members from the College sharing their experiences in implementing curriculum changes and coordinating these changes across courses. The majority of the workshop was dedicated to a case study activity in which participants simulated the process of building a new course. This activity was adapted from a Project Kaleidoscope (PKAL) case study created by Daniel F. Sullivan on how to revise course curriculum (www.pkal.org). The 90-min workshop is outlined below and described in the pages that follow.

1. **Faculty Experience and Research (30 min)**
2. **Case Study—Lark University**

   (a) **Part I: Discussing the proposed curriculum change (25 min)**
   (b) **Part II: Implementing the curriculum change (25 min)**

3. **Closure and Open Conversation (10 min)**

1. **Faculty Experience and Research**

Two groups of faculty members shared examples of how they have worked with their peers who teach connected courses to change curriculum across these courses. The presenters shared different facets of the curriculum change process and how the collaboration with their peers supported student learning progressions across courses.

Introductory Chemistry Courses: The first group discussed the initiative undertaken by the Chemistry and Biochemistry department to change the sequence of introductory general and organic chemistry courses. They described key aspects of the curriculum change process, including creating a task force, looking at effective practices in other institutions, and involving the entire departmental or College community in collaborative decision-making. The presenters also shared important considerations in weighing potential curriculum changes. These considerations include implications in terms of the GTA work force, availability of lab spaces, and the availability and cost of textbooks appropriate for the proposed sequencing.

Sequence of Undergraduate Courses with a Focus on Host Pathogen Interactions: The second set of presenters shared how a group of faculty members with shared research on Host Pathogen Interaction (HPI) collaborated to change the curriculum across a sequence of nine undergraduate courses that they teach. The presenters explained that their overarching goals were to create bridges between the nine courses, eliminate excessive overlap across courses, and structure curriculum such that concepts and ideas introduced in one course become the foundation for concept development in successive courses. The presenters also discussed how, in order to develop such a continuum between the courses, the HPI community not only worked on curriculum but also on teaching approaches and rigorous assessment to promote concept development. To measure students' concept understanding, the group developed the HPI Concept Inventory that centers on a set of 13 concepts considered fundamental to the understanding of HPI (Marbach-Ad et al., 2007, 2009, 2010).

2. Role Play/Case Study—Lark University

(a) Part I: Discussing the proposed curriculum change

Workshop participants split into groups of approximately seven to engage in a case study about a proposed curriculum change initiative in the fictional Lark University (see Case Study in box).

Lark University Case Study

Lark University decided to put increasing emphasis on excellence in teaching and learning in all sciences. The University strives to be a world leader in science education among research intensive universities. Therefore, they would like to completely overhaul the undergraduate science curriculum. The dean is looking for proposals. There is a limited amount of money and all the science departments can compete for the funds. Dr. Sparrow decided that he wants to submit a proposal to change the SCI 381 undergraduate course called "Transforming the transform."

Dr. Sparrow will teach SCI 381 for the first time next semester. He received the textbook from last year, the old syllabus (which hasn't changed for the last ten years), and the old exams. The old course format was three 50-min lecture periods, one 50-min discussion session, and one three-hour lab per week.

Dr. Sparrow will teach his version of the SCI 381 course in parallel to Dr. Loon, who taught the course last year. Dr. Loon reported that over the past five years students have complained that the course lectures were not interesting and the material was way over their heads. The course has a prerequisite of SCI 212 (called "Those who transform-transform") that Dr. Falcon has taught for several years.

A group is meeting to discuss submitting a curriculum proposal to the dean. Around the table are Dr. Sparrow, Dr. Loon, Dr. Falcon, the SCI 381 discussion session GTA (Beak Humming), an undergraduate student who took the course last year (Wing Stork), the lab coordinator (Dr. Feather) and the department chair (Dr. Osprey).

Adapted from PKAL Loon University Case Study, Daniel F. Sullivan (www. pkal.org/documents/Vol4LoonUniversity.cfm)

After familiarizing themselves with the case, participants discussed the proposed initiative, with each acting as one of the seven key players: the instructor who sought to change the course, two instructors of related courses, the GTA and lab coordinator associated with the course, an undergraduate student who had previously taken the course, and the department chair. This discussion revolved around three central questions:

1. What things do you need to consider in changing a course?
2. Who should be involved in making changes to the course?
3. What will be the role of each key player?

Following the small group activity, each group reported how they responded to the three questions.

(b) **Part II: Implementing the curriculum change**

The small groups then reconvened and each group was instructed to think through the process of implementing the course change after the suggested initiative received funding. They discussed the steps that the course change would require, priorities in this process, and the sequence of change. To guide this discussion, each group was given the following prompt:

Good news—you got the funds!!! Now is the time to start making changes—but you teach the course next semester and not much time is left. So you think of going step by step:

1. What would be your top priority?
2. What would you change first?

3. **Closure and Open Conversation**

The entire group reconvened to summarize each small group discussion. The facilitator summarized commonalities and noted significant differences across groups. She emphasized that curriculum change requires coordination across courses and key players, and that the process should be well planned, regularly assessed, and open to revision. She then opened the floor to conversation and invited participants to suggest ways in which the College could improve curriculum and instruction.

PCK 3: Blended Learning: What It Is and How to Use It

In this seminar, four panelists from across the university shared how they had successfully incorporated blended learning techniques into their own classes, and discussed their challenges and successes in doing so. The 90-min seminar is outlined below.

1. **Introduction and Definition of Key Terms (10 min)**
2. **Faculty Experience with Blended Learning (40 min)**
3. **Questions for the Panel (20 min)**
4. **Open Conversation (20 min)**

1. Introduction and Definition of Key Terms

At the start of the seminar, the facilitator provided a definition of blended learning and an overview of what constitutes blended learning.

2. Faculty Experience with Blended Learning

The panel consisted of four presenters who shared their experiences in planning and using a blended learning approach. Because of the diverse classes that they taught, each panelist described the context of their courses as well as how blended learning fit into that context.

The first panelist, a professor in the Department of Kinesiology, described how he flipped his large enrollment class (about 200 students) using blended learning techniques. While his previous model consisted of holding two face-to-face lecture sessions and one discussion session per week, the new model included one face-to-face lecture session, one discussion session, and one online lecture per week. In this new format, students were responsible for learning more content outside of class and had access to a variety of supplemental resources (e.g., recorded lectures, podcasts, a curated list of relevant papers and websites, and optional course textbooks). Before each face-to-face lecture, students took an online quiz that served as a formative assessment to give them and the instructor feedback on their understanding of the content. This also incentivized students to learn the material. The lecture session was then devoted to discussion of key concepts and principles, with a focus on any concepts that students had difficulty with, as indicated by the quizzes. Flipping the responsibility for learning basic content to outside of class created time for activities and inquiry-based learning during face-to-face class meetings.

The second and third panelists described how they incorporated blended learning into a molecular biology course enrolling about 60 students. Each of the panelists teaches this course independently, in different semesters. The second panelist described their process for collaboratively defining what content should be covered in the course and how to best distribute that content between face-to-face and online modes of instruction. She also talked about intramural funding sources that support transitions to blended learning. The third panelist shared his experience implementing this blended learning approach. He talked about the pros and cons of the new approach, and the lessons he learned in the process. He discussed how the course content, molecular biology, was well suited to blended learning because of the wide range of high quality, online resources that are available. In addition the field of molecular biology is moving to an increasing reliance on simulations, models, visualizations, and database tools that are better taught through online and multimedia instruction rather than textbooks and lectures.

The fourth panelist described how he used online video modules to supplement his teaching in a large enrollment (about 200 student) introductory chemistry class. He developed a series of 12 videos demonstrating how to solve representative problems from the course textbook. The videos were created using audio and screen capture software and posted on the course website. He explained how he evaluated the use and effectiveness of the videos through statistical analysis of student quiz and exam scores.

3. Questions for the Panel

Following the four presentations, we asked all panelists a series of questions:

1. What are the major components for successful blended learning that includes online teaching (e.g., class size, student profile, level of interactivity, assessment, instructor involvement, level of feedback)?
2. How did you decide which components of the class to teach through face-to-face and which components to teach through online instruction?
3. How does the content or pedagogical approach differ in each teaching method?
4. How do you integrate what is learned online into classroom instruction?
5. How have you fostered collaborative learning in online communities? What challenges have you faced in doing so?
6. What challenges have you faced in transitioning from traditional instruction to blended learning?
7. How do you assess the effectiveness of your online instruction?
8. How do you integrate assessment of the effectiveness of face-to-face and online instruction?

4. Open Conversation

The seminar ended with a 20-minute question and answer period. Several of those in the audience were already incorporating some elements of blended learning in the courses they taught, and they shared their experiences and asked for advice from the panel.

PCK 4: As You Can See, But Students Don't: A Closer Look at How Students Read and Interpret Graphic Information

This workshop was led by Dr. Virginia Anderson, who was then Professor of Biology at Towson University. Dr. Anderson is an expert in assessment in science and is the co-author of the book *Effective Grading: A Tool For Learning and Assessment in College* (Walvoord & Anderson, 2010). She provided a brief overview of student assessment to serve as a foundation for the hands-on activities that followed. These activities allowed participants to build their skills in creating or evaluating assessment tools and analyzing student responses on those tools. The workshop was structured as follows:

1. **Introduction and the Importance of Assessment (20 min)**
2. **Using Different Types of Graphs and Charts as Assessment Tools (70 min)**

1. Introduction and the Importance of Assessment

Dr. Anderson described the role and purposes of course assessment, how course assessment can be used to measure different levels of thinking (Bloom, 1984), and strategies to increase the accuracy and authenticity of assessment. She discussed how assessment tools can be better integrated with course content to provide valuable information for instructors as well as students. She emphasized that students take their cues as to what is important in a course from the types of assessments and the value assigned to each by the instructor. If assessments measure rote learning of disconnected facts, students will generally not be motivated to exercise higher order cognitive skills and, as a result, may face difficulties in subsequent courses or in the workplace when they are expected to think synthetically and critically.

2. Using Different Types of Graphs and Charts as Assessment Tools

After providing this brief overview, Dr. Anderson led a hands-on activity in which participants split into small groups to discuss strategies for helping students read and interpret graphic information. They also examined sample assessment tools and rubrics for assessing students' knowledge and skills in this area, evaluated sample student responses to identify where students had gaps in their knowledge, and discussed strategies for addressing these gaps.

> **Recommended Resources**
> In the seminar, Anderson included material from her book, *Effective Grading: A Tool for Learning and Assessment in College* (Walvoord & Anderson, 2010), and suggested that participants could find additional examples in *Assessing Student Learning: A Common Sense Guide* (Suskie, 2010). Much like her seminar, these resources include examples and sample tools that are intended specifically for undergraduate science courses, making them particularly relevant to our audience.

PCK 5: Identifying the Hallmarks of Effective Teaching to Create a Rubric for Peer Review of Teaching

This workshop occurred as a part of an initiative to develop and implement a new system for the peer review of teaching in the biology department. It was given as part of a mandatory faculty retreat, and all faculty members within the department were expected to participate. The workshop was led by the department chair, a peer review committee established by the chair shortly before the retreat, and TLC staff. In the weeks leading up to the workshop, the peer review committee had studied peer review processes at other universities to develop a sense of the range of possible outcomes. The workshop provided professional development for all participants in the form of conversations about what constitutes effective teaching. These conversations engendered a shared sense of ownership of the peer review process and an understanding of this process as arising from and connected to the departmental culture.

The workshop included multiple whole-group and small-group components:

1. **Pre-Workshop Reading (Independent)**
2. **Determining Goals of the Peer Review Process (70 min)**
3. **Defining Hallmarks of Effective Teaching (60 min)**
4. **Using Rubrics in the Peer Review Process (20 min)**
5. **Measuring the Hallmarks of Effective Teaching (50 min)**
6. **Developing Peer Review Process and Procedures (10 min)**

1. **Pre-Workshop Reading**

Prior to the workshop, all participants were given an annotated bibliography that summarized findings from a review of literature on the peer review process. These materials offered participants an introduction to the different ways in which peer reviews can be conducted and highlighted some peer review systems particularly appropriate for the context of a university science department. The readings also identified common goals of the peer review process, such as the need for the process to be of value not only to the faculty members who are being reviewed but also to the faculty members conducting the reviews.

Peer Review of Teaching Literature Review: Key Points

- Training in classroom observation can increase the reliability of peer classroom observation (Paulson, 2002).
- The most trustworthy observers are those who know the disciplinary content of the course being reviewed (Yon, Burnap, & Kohut, 2002).
- Peer ratings of teaching performance and materials are complimentary to student ratings in creating a very comprehensive picture of teaching effectiveness. Peer ratings also cover aspects of teaching that students are not in a position to evaluate (Berk, 2005).
- Both observers and observees value the peer observation process and believe that peer observation reports are valid and useful (Kohut, Burnap, & Yon, 2007).
- Peer review puts faculty in charge of the quality of their work as teachers (Hutchings, 1996).
- Peer reviews should not be viewed as punitive; instead, they should facilitate reflection on teaching styles, strategies, and philosophies for the benefit of increased student learning (Blackmore, 2005).

2. **Determining Goals of the Peer Review Process**

The department chair introduced the workshop and described the goal of creating a peer review process that would be sustainable and would contribute to the improvement of teaching within the department. Then, participants split into groups of three to five to consider the goals of the peer review process. After approximately 30 min of small group discussion, each group reported out their goals. The workshop facilitator created a single list of all unique goals on the board. Each participant then identified the three goals that s/he viewed as most important by placing a sticky dot next the respective goals. At the end of this section of the workshop, the facilitator summarized the goals with broad faculty support:

1. Improve teaching effectiveness of individual faculty by providing feedback on teaching
2. Expose reviewers to a range of teaching styles and approaches, and thereby make them more aware of best practices
3. Provide information that can be used to assess and adjust, if necessary, course content for the audience or curriculum
4. Provide data for accurate and equitable decisions on tenure, promotion, and merit pay increases

3. **Defining Hallmarks of Effective Teaching**

In this component of the workshop, participants discussed and prioritized the characteristics of effective teaching. Workshop participants again split into small groups, with the groups re-mixed from the earlier group activity to facilitate transfer of ideas between groups and allow faculty to hear the thoughts of a large number of colleagues. The format of this component was very similar to the previous component. Each group had an opportunity to create their own list of the hallmarks of effective teaching, and then reported out to the entire group while the facilitator created a list of all unique hallmarks. Again, participants used dot voting to indicate the three most important characteristics of effective teaching. Based on the results of the dot voting, the facilitator created a rank order list of the hallmarks and highlighted the twelve hallmarks receiving the most votes:

1. Demonstrate enthusiasm to inspire students
2. Engage students through active learning activities
3. Challenge students
4. Get students to think like scientists
5. Relate course material to everyday life, research, and other courses to make it more relevant to students
6. Make learning outcomes and expectations clear
7. Promote math and quantitative skills
8. Promote reading and writing
9. Revisit core concepts throughout the course and build themes across lectures
10. Provide opportunities for students to practice important skills
11. Inspire students to want to learn more about course material
12. Provide feedback to students throughout the course

4. **Using Rubrics in the Peer Review Process**

The TLC director provided a brief overview of rubrics and how they can be used in the peer review process. She defined a rubric as a tool for communicating expectations about quality. A rubric provides a set of criteria and standards that are typically linked to objectives that are used to assess performance. She also discussed key elements of a rubric and considerations for developing one.

Pros and Cons of Using Rubrics for Peer Evaluation
Pros

- Rubrics can increase the assessee's understanding of the task and the expectations about quality.
- Rubrics can facilitate the work of the assessor and promote accurate and fair evaluation on the designated criteria.
- Rubrics can elicit quantitative and qualitative feedback for the assessee and the assessor.
- Rubrics can highlight salient strengths and weaknesses, and may also point assessees to areas in which they can improve their teaching practices.

Cons

- A rubric can diminish the breadth or depth of the evaluation process.
- An inappropriately designed rubric can be more detrimental than beneficial if it does not emphasize the desired/key elements that are being evaluated.
- The possibility that rubrics may be used beyond their original purpose (e.g., using rubrics that were designed for formative review to make high stakes promotion decisions) can create apprehension or other negative feelings about the rubric and related processes.

5. Measuring the Hallmarks of Effective Teaching

Workshop participants split into four groups, with participants again mixing to form new groups. Each of these groups was assigned three of the top 12 hallmarks of effective teaching, and asked to brainstorm about methods to measure each hallmark. Each group then reported out their top three methods for measuring each hallmark. The facilitator again created a master list, and participants were invited to recommend additional methods.

6. Developing Peer Review Process and Procedures

To open the discussion of what kind of process should be instituted in the department, the department chair presented an overview of processes and procedures for peer review at other universities. He raised four questions that the department should address in developing its own peer review process:

1. Who should conduct the peer reviews?
2. How often should each faculty member be reviewed?
3. How many classroom visits should the review entail?
4. Should reviewer identity be kept confidential?

All participants completed a survey to vote on their preferred process and procedures.

Following the Workshop

Over the next few months, the peer review committee reconvened to review feedback from the workshop and the surveys, and to finalize the development of the peer review rubric. This rubric included many of the hallmarks of effective teaching that were identified during the workshop, and also drew heavily on suggested methods for measuring these hallmarks. The group also developed procedures for the peer review process, and in doing so incorporated the majority responses from the survey.

The peer review process was piloted over the course of a year, and faculty had the opportunity to review and provide feedback on the observation rubric. To further validate the rubric, faculty observations were compared to student end-of-semester evaluations. The rubric and procedures were revised based on the pilot and are now ready for full implementation (see Peer Review Rubric and Rubric Summary Report in supplemental materials).

Feedback on the Workshop and the Peer Review Process

Biology faculty members provided positive feedback about the workshop as well as the peer review process. Some faculty had initially questioned whether this workshop deserved three hours of their time and if developing a rubric for peer review was an appropriate activity for a departmental retreat. Afterwards, many faculty members commented that the workshop served as professional development in itself. They stated that the whole activity was relevant to them and required them to engage intellectually with topics that are generally not discussed in faculty meetings. Faculty members also provided positive feedback about the peer review process itself. Many faculty members noted that they learned a lot, not just from being observed and receiving feedback on their teaching, but also through the act of observing others and applying the rubric.

References

Angelo, T., & Cross, P. (1993). *Classroom assessment techniques: A handbook for college teachers* (2nd ed.). San Francisco, CA: Jossey-Bass.

American Association for the Advancement of Science (AAAS). (2011). *Vision and change: A call to action*. Washington, DC: AAAS.

Ausubel, D. (1968). *Educational psychology: A cognitive view*. New York, NY: Rinehart & Winston.

Avargil, S., Herscovitz, O., & Dori, Y. J. (2012). Teaching thinking skills in context-based learning: Teachers' challenges and assessment knowledge. *Journal of Science Education and Technology, 21*, 207–225.

Berk, R. A. (2005). Survey of 12 strategies to measure teaching effectiveness. *International Journal of Teaching and Learning in Higher Education, 17*, 48–62.

Birenbaum, M. E., & Dochy, F. J. (1996). *Alternatives in assessment of achievements, learning processes and prior knowledge*. Boston, MA: Kluwer Academic Publishers.

Black, P. J., & Wiliam, D. (1998). Assessment and classroom learning. *Assessment in Education: Principles Policy and Practice, 5*(1), 7–73.

Blackmore, J. A. (2005). A critical evaluation of peer review via teaching observation within higher education. *International Journal of Educational Management, 19*, 218–232.

Bloom, B. S. (1976). *Human characteristics and school learning*. New York, NY: McGraw-Hill.

Bloom, B. S. (1984). *Taxonomy of educational objectives. Handbook 1: Cognitive domain*. New York, NY: Longman.

Brandwein, P. F. (1977). What is curriculum? *National Elementary Principal, 57*(1), 10–11.

Bruner, J. (1960). *The process of education*. Cambridge, MA: Harvard University Press.

Bush, S. D., Pelaez, N. J., Rudd, J. A., Stevens, M. T., Tanner, K. D., & Williams, K. S. (2008). Science faculty with education specialties. *Science, 322*, 1795–1796.

Cathcart, L. A., Stieff, M., Marbach-Ad, G., Smith, A. C., & Frauwirth, K. A. (2010). *Using knowledge space theory to analyze concept maps*. Paper presented at the 9th international conference of the learning sciences.

Dochy, F., Segers, M., & Buehl, M. M. (1999). The relation between assessment practices and outcomes of studies: The case of research on prior knowledge. *Review of Educational Research, 69*, 145–186.

Duncan, R. G., & Rivet, A. E. (2013). Science education. Science learning progressions. *Science, 339*(6118), 396–397. doi:10.1126/science.1228692.

Duschl, R. A., Schweingruber, H. A., & Shouse, A. W. (2007). *Taking science to school: Learning and teaching science in Grades K-8*. Washington, DC: National Academies Press.

Ebert-May, D., Derting, T. L., Hodder, J., Momsen, J. L., Long, T. M., & Jardeleza, S. E. (2011). What we say is not what we do: Effective evaluation of faculty professional development programs. *Biosciene, 61*(7), 550–558.

Ebert-May, D., & Hodder, J. (2008). *Pathways to scientific teaching*. Sunderland, CT: Sinauer Associates, Inc.

Egan, K. (2003). What is curriculum? *Journal of the Canadian Association for Curriculum Studies, 1*(1), 9–16.

Fisher, K. M. (1983). *Amino acids and translation: a misconception in biology*. Paper presented at the International Seminar on Misconceptions in Science and Mathematics, Cornell University, Ithaca, NY.

Fortus, D., & Krajcik, J. (2012). Curriculum coherence and learning progressions. In B. J. Fraser, K. T. Kenneth, & C. J. McRobbie (Eds.), *Second international handbook of science education* (pp. 783–798). Dordrecht, The Netherlands: Springer.

Freeman, S., Eddy, S. L., McDonough, M., Smith, M. K., Okoroafor, N., Jordt, H., & Wenderoth, M. P. (2014). Active learning increases student performance in science, engineering, and mathematics. *Proceedings of National Academic Science U S A 111*(23), 8410–8415. doi: 10.1073/pnas.1319030111

Freeman, S., O'Connor, E., Parks, J. W., Cunningham, M., Hurley, D., Haak, D., . . . Wenderoth, M. P. (2007). Prescribed active learning increases performance in introductory biology. *CBE Life Sciences Education, 6*, 132–139.

Gilbert, J. K., Osborne, R. J., & Fensham, P. J. (1982). Children's science and its consequences for teaching. *Science Education, 66*(4), 623–633.

Graff, N. (2011). An effective and agonizing way to learn: Backwards design and new teachers' preparation for planning curriculum. *Teacher Education Quarterly, 38*(3), 151–168.

Handelsman, J., Ebert-May, D., Beichner, R., Bruns, P., Chang, A., DeHaan, R., . . . Wood, W. B. (2004). Scientific teaching. *Science, 304*(5670), 521–522.

Handelsman, J., Miller, S., & Pfund, C. (2007). *Scientific teaching.* New York, NY/Englewood, CO: W.H. Freeman & Company in collaboration with Roberts & Company Publishers.

Henderson, C., Beach, A., & Finkelstein, N. (2011). Facilitating change in undergraduate STEM instructional practices: An analytic review of the literature. *Journal of Research in Science Teaching, 48*(8), 952–984.

Hutchings, P. (1996). The peer review of teaching: Progress, issues and prospects. *Innovative Higher Education, 20,* 221–234.

Kauffman, D., Moore, J. S., Kardos, S., Liu, E., & Peske, H. (2002). Lost at sea: New teachers' experiences with curriculum and assessment. *The Teachers College Record, 104*(2), 273–300.

Kember, D., & Kwan, K. P. (2000). Lecturers' approaches to teaching and their relationship to conceptions of good teaching. *Instructional Science, 28*(5), 469–490.

Knowles, M. S. (1980). *The modern practice of adult education: Andragogy versus pedagogy.* Englewood Cliffs, NJ/Cambridge, UK: Prentice Hall.

Kohut, G. F., Burnap, C., & Yon, M. G. (2007). Peer observation of teaching: Perceptions of the observer and the observed. *College Teaching, 55*(1), 19–25.

Krause, S. J., Kelly, A., Tasooji, J., Corkins, B. D., & Purzer, S. (2010). Effect of pedagogy on conceptual change in an introductory materials science course. *International Journal of Engineering Education, 26,* 869–879.

Magnusson, S., Krajcik, L., & Borko, H. (1999). Nature, sources and development of pedagogical content knowledge. In J. Gess-Newsome & N. G. Lederman (Eds.), *Examining pedagogical content knowledge* (pp. 95–132). Dordrecht, The Netherlands: Kluwer.

Marbach-Ad, G. (2009). From misconceptions to concept inventories. *Focus on Microbiology Education, 15*(2), 4–6.

Marbach-Ad, G., Briken, V., El-Sayed, N., Frauwirth, K., Fredericksen, B., Hutcheson, S., . . . Smith, A. C. (2009). Assessing student understanding of host pathogen interactions using a concept inventory. *Journal of Microbiology and Biology Education, 10,* 43–50.

Marbach-Ad, G., Briken, V., Frauwirth, K., Gao, L. Y., Hutcheson, S. W., Joseph, S. W., . . . Smith, A. C. (2007). A faculty team works to create content linkages among various courses to increase meaningful learning of targeted concepts of microbiology. *CBE life sciences education, 6*(2), 155–162.

Marbach-Ad, G., McAdams, K., Benson, S., Briken, V., Cathcart, L., Chase, M., . . . Smith, A. (2010). A model for using a concept inventory as a tool for students' assessment and faculty professional development. *CBE Life Science Education, 9,* 408–436.

Marbach-Ad, G., & Sokolove, P. G. (2000). Can undergraduate biology students learn to ask higher level questions? *Journal of Research in Science Teaching, 37*(8), 854–870.

Marbach-Ad, G., & Stavy, R. (2000). Students' cellular and molecular explanations of genetic phenomena. *Journal of Biological Education, 34*(4), 200–205.

Marzano, R. J. (2004). *Building background knowledge for academic achievement: Research on what works in schools.* Alexandria, VA: ASCD.

Middle States Commission on Higher Education. (2007). Student learning assessment: Options and resources (2nd ed.). Philadelphia, PA: Middle States Commission on Higher Education.

Mintzes, J. J., Wandersee, J. H., & Novak, J. D. (2005). *Teaching science for understanding: A human constructivist view.* San Diego, CA: Academic Press: An Elsevier Imprint.

Novak, J. D., & Gowin, D. B. (1984). *Learning how to learn.* London, UK: Cambridge University Press.

NRC. (2003). *National Research Council, Bio 2010: Transforming undergraduate education for future research biologists.* Washington, DC: National Academy Press.

Paulson, M. B. (2002). Evaluating teaching performance. *New Directions for Institutional Research, 114,* 5–18.

Piaget, J. (1954). *The construction of reality in the child.* New York, NY: Routledge.

Sadler, D. R. (1989). Formative assessment and the design of instructional strategies. *Instructional Science, 18,* 119–144.

Schneps, M. H., & Sadler, P. M. (Writers). (2003). A private universe. Minds of our own. Available in the form of DVD, an electronic source.

Shulman, L. S. (1986). Those who understand: Knowledge growth in teaching. *Educational Researcher, 15*(2), 4–31.

Smith, M. K., Wood, W. B., Krauter, K., & Knight, J. K. (2011). Combining peer discussion with instructor explanation increases student learning from in-class concept questions. *CBE—Life Sciences Education, 10*, 55–63.

Stavy, R., & Tirosh, D. (2000). *How students (mis-) understand science and mathematics: Intuitive rules*. New York, NY: Teachers College.

Stieff, M. (2005). Connected chemistry—A novel modeling environment for the chemistry classroom. *Journal of Chemical Education, 82*(3), 489–493.

Suskie, L. (2010). *Assessing student learning: A common sense guide*. Hoboken, NJ: John Wiley & Sons.

Thijs, G. D., & van den Berg, E. (1993, August 1–4). *Cultural factors in the origin and remediation of alternative conceptions*. Paper presented at the third international seminar on misconceptions and educational strategies in science and mathematics, Cornell University, Ithaca, NY.

Tobias, S. (1994). Interest, prior knowledge and learning. *Review of Educational Research, 64*(1), 37–54.

Vygotsky, L. S. (1978). *Mind in society: The development of higher psychological processes*. Cambridge, MA: Harvard University Press.

Walvoord, B. E., & Anderson, V. J. (2010). *Effective grading: A tool for learning and assessment in college* (2nd ed.). San Francisco, CA: Jossey-Bass.

Wieman, C. (2007). Why not try a scientific approach to science education? *Change*.http://www.changemag.org/Archives/Back%20Issues/September-October%202007/index.html

Wiggins, G. P., & McTighe, J. (1998). *Understanding by design*. Alexandria, VA: Association of Supervision and Curriculum Development.

Wyse, S. A., Long, T. M., & Ebert-May, D. (2014). Teaching assistant professional development in biology: Designed for and driven by multidimensional data. *CBE Life Science Education, 13*, 212–223.

Yon, M., Burnap, C., & Kohut, G. (2002). Evidence of effective teaching perceptions of peer reviewers. *College Teaching, 50*, 104–110.

Chapter 3
Acculturation of New Faculty Members

Teaching for the first time can be a daunting process... In a way,
the first class becomes an experimental class.

–Robert, Chemistry member

New faculty members come to their jobs with diverse backgrounds and experiences, but typically little formal preparation for teaching. They often have a wide variety of concerns, such as limited teaching experience and effectively managing their time given their multiple responsibilities. Based on a longitudinal study that included pre- and post-interviews with new faculty members, the Teaching and Learning Center (TLC) has tailored many professional development activities to meet the special needs of new faculty members. Through these activities, new faculty members are integrated into the departmental community and learn the skills they need to be more effective instructors. In this chapter, we provide an overview of the preparation and work context of new faculty members, elaborate on the findings from our longitudinal study, and describe our professional development program for new faculty members.

Preparation and Work Context of New Faculty Members

One challenge many new faculty members face is adjusting to their many responsibilities, which generally include research, teaching, service, and mentoring (Boice, 2000; Reybold, 2003). Many new faculty members report that they find it difficult to identify the appropriate balance between these different responsibilities (Austin, Sorcinelli, & McDaniels, 2007). Furthermore, these responsibilities are not equally emphasized by colleges and universities, which often send an implicit message that prioritizes success in research over success in teaching.

This implicit message to prioritize research over teaching begins in graduate school. Graduate programs provide extensive training to prepare graduate students for their research responsibilities, but typically offer little formal preparation for teaching (Cox, 1995; Golde & Dore, 2001; Handelsman, Miller, & Pfund, 2007; Luft, Kurdziel, Roehrig, & Turner, 2004). In some instances, graduate students may

© Springer International Publishing Switzerland 2015   55
G. Marbach-Ad et al., *A Discipline-Based Teaching and Learning Center*,
DOI 10.1007/978-3-319-01652-8_3

be advised to devote minimal time to developing their teaching skills so that they can focus primarily on research endeavors (Austin & McDaniels, 2006; Golde & Dore, 2001). The relative lack of emphasis on training in teaching indicates to future faculty that teaching will not be valued as much as research in their academic careers (Austin, 2002; Nyquist et al., 1999).

In their first faculty position, many faculty members report experiencing a myriad of difficulties related to fulfilling their teaching responsibilities. These difficulties include excessive worrying about their teaching, unclear expectations, spending a disproportionate amount of time preparing for lectures, classroom incivilities, and classroom management issues (Boice, 2011).

Institutional support to assist faculty in their teaching varies across universities and is greatly influenced by the priorities of the institution (Austin & McDaniels, 2006; Bouwma-Gearhart & Schmid, 2012; Boyer Commission on Educating Undergraduates in the Research University, 1998). In research universities, many incoming faculty members have ample opportunities to collaborate with university colleagues on research projects. However, they rarely have similar opportunities to collaborate on teaching endeavors. Instead, faculty members work in relative isolation as they select course content and decide upon pedagogical approaches (Tanner & Allen, 2002). Some faculty members, especially new faculty in tenure-track positions, are disinclined to ask their departmental peers for help because they fear that needing help could be seen as a weakness and have negative repercussions for achieving tenure and promotion (Boice, 2011). This issue is exacerbated in settings in which new faculty are not introduced to formal professional networks and communities to support them as they adjust to their teaching responsibilities and expectations.

Professional development programs and institutional support can ease this transition and foster the development of faculty as effective teachers (Austin et al., 2007). Reaching this population can have long-lasting impact. Most new faculty members do not yet have years of experience that can lead to deeply ingrained teaching practices, so they may be more open to adopting innovative approaches. Furthermore, this support could cultivate the next generation of change agents that can influence the departmental culture around teaching.

Getting to Know Our New Faculty Members

When the TLC began, we knew that we needed to provide new faculty with targeted support. Our initial support for new faculty was a welcome workshop that began as an initiative of the College Dean's Office. These annual workshops, held early in the fall semester, brought together all newly arrived faculty members to discuss a variety of topics related to teaching and learning. New faculty members were also given welcome packets that included influential literature from the science education field. In addition to providing new faculty with essential information and resources, the workshops also served to build community amongst the new faculty members and created a connection between them and the TLC. Workshop details and materials can be found in the Implementation Guide at the end of this chapter.

In order to customize more fully the support we offered to new faculty members, we decided to interview one cohort of incoming faculty members to better understand what they needed in the context of our College. To this end, we conducted a longitudinal study that involved in-depth interviews in their first and third years at the university. This study was valuable not only in informing our professional development, but also as professional development in itself, as it required the faculty members to reflect on their teaching, consider their strengths and weaknesses, and review steps that they had undertaken to improve their teaching and student learning. The interviews also strengthened the connection between the new faculty members and the TLC staff who conducted the interviews.

Overview of the New Faculty Longitudinal Study

We followed 11 new biology and chemistry faculty members over the course of their first three years at the University (2007–2010). This cohort of new faculty included five females and six males, nine of whom were hired in tenure-track positions (with research and teaching responsibilities) and two hired as non-tenure-track lecturers (with only teaching responsibilities). All new faculty members reported that they had gained some teaching experience previously as graduate students or postdoctoral fellows. This experience generally consisted of supervising laboratory courses or leading discussion sections. Four new faculty members reported that they did not have any previous experience teaching a lecture class. Those who came to the University with substantive previous teaching experience had taught at various levels, including graduate medical student courses (three faculty members), undergraduate courses (two faculty members), and high school courses (two faculty members).

The Cohort of New Faculty Members (2007)*

Dana and **Amber** are females, and were hired as biology lecturers.

Jenna, **Susan**, and **Linda** are females, and were hired as assistant professors in biology.

Ryan and **John** are males, and were hired as assistant professors in biology.

Robert is a male, and was hired as an assistant professor in chemistry.

David and **Tim** are males, and were hired as associate professors in biology.

Scott is a male, and was hired as an associate professor in chemistry.

*To protect the identities of study participants, all names have been changed.

This longitudinal study shed light on the diverse types of experience possessed by incoming faculty members, as well as the challenges they faced in adapting to their new teaching responsibilities and how they addressed these challenges. The interview protocols that guided the initial and follow-up interviews were designed to encompass pedagogical content knowledge (PCK) theory and its five constituent components. Chapter 1 of this volume contains a more detailed discussion of PCK. Marbach-Ad, Schaefer Ziemer, Thompson, and Orgler (2013) provides a detailed account of findings from this longitudinal study.

Analysis of Study Findings by PCK Component

This section provides a detailed description of faculty responses to the interview questions, organized by PCK component. For each question, we highlight major themes that emerged across study participant responses. We provide quotations from the participating faculty members to illustrate these themes, but use pseudonyms to protect their anonymity.

PCK 1: Student Understanding of Science

The first PCK component relates to students' prior knowledge, their alternative conceptions, how they build scientific knowledge, and how student diversity impacts learning. Below, we elaborate on participant responses related to how they learned about students' background knowledge in order to relate to the diversity of the students in their classroom.

In the initial interview, only four of the eleven new faculty members reported having ways of learning about their students' backgrounds. The remaining seven did not yet have a formalized way of learning about their students' backgrounds. As David, explained, "not having done it, I do not know how."

In the second interview (three years later), nine of the eleven faculty members had established methods to build a comprehensive picture of their students' prior knowledge and academic goals, interests, and preferences. Faculty members often used multiple complementary methods to learn about their students' background. These methods differed based on the course and the instructor. Commonly reported methods included:

1. Investigating content coverage in prior courses by talking with instructors of those courses
2. Measuring content knowledge at the start of the course

   - Ungraded pretest
   - Graded exam
   - Clicker questions

3. Learning about the composition of the class

- Campus database
- Survey
- Introductions

4. Measuring learning styles and attitudes

- Survey of attitudes towards science and research
- Survey of learning styles

On the pages that follow, we describe how the new faculty members used these methods.

1. Investigating Content Coverage in Prior Courses

Six faculty members sought to gain an understanding of what was taught in prerequisite courses or courses generally taken prior to their course. Some looked at syllabi and/or textbooks, while others spoke to the instructors responsible for these courses about "what they actually cover."

2. Measuring Content Knowledge at the Start of the Course

Three faculty members gave students a diagnostic test to measure their prior knowledge. Information from this test was often supplemented by speaking with the instructors of prior courses. Tim explained why he used both methods: "They clearly should know [material from the courses they have taken]. Now, whether they remembered it, [that is] a different story." In order to measure what students retained from prerequisite courses, he incorporated "the same exact questions" from the prerequisite courses' final exam into his diagnostic test.

In courses that require interdisciplinary knowledge, some faculty members emphasized the need to understand students' relevant background knowledge from other disciplines. As Scott explained:

> At the beginning [of the chemistry course], I will give a math quiz—not for grades, but just to see where they are mathematically. Because in the class that I've been teaching, the biggest hurdle is that some of the students don't have the proper math background.

Faculty members sometimes used the first graded exam to assess students' background knowledge. In these cases, they offered their first exam during the 10-day schedule adjustment period to help students decide whether they had sufficient background knowledge to continue in the course. Tim reported that the first exam in his course covered "what you should know going in . . . If you're struggling here, you should strongly consider dropping the class because you will struggle later." Some faculty members reported using less formal mechanisms. For example, Amber reported questioning the entire class using clickers to gauge whether students were generally familiar with key topics from prior classes.

3. Learning About the Composition of the Class

Faculty members used different methods to learn about the composition of the class in terms of majors, year, and, in some cases, relevant prior coursework that may not be an explicit pre-requisite for the course. Five faculty members relied on the campus student information system for this purpose. David, who taught an upper-level interdisciplinary class with students coming from multiple departments, reported that he found it important to view "the various descriptive statistics" provided by the campus database to learn about students' majors and prior coursework. As David explained:

> I made correlations between what they've taken before they come to my class and how they perform. And it has helped me quite a bit. I know that we don't require [name of course] as a prerequisite, but people who take this course before coming to my class do much better.

Two faculty members asked the students to provide information about their background during the first class session. David asked students to introduce themselves and share "what they've done... what research they're doing." Jenna had students complete a short survey to gather information on their major, prior coursework, and plans after college.

4. Measuring Learning Styles and Attitudes

Two faculty members gathered information about students' attitudes and learning styles. Amber surveyed student attitudes towards science by asking questions such as "if they like doing research." Jenna used two surveys to both "get a little bit of a feel for what they're interested in" in terms of learning science and to understand their learning style. For understanding learning styles, she used the Index of Learning Style Questionnaire (Soloman and Felder, www.engr.ncsu.edu/learningstyles/ilsweb.html), which is one of the resources that is provided by the TLC to new faculty in their welcome packet and discussed in the new faculty welcome workshop. Jenna used this validated instrument to characterize the diversity of learning styles represented in the classroom and adjusted her teaching and content delivery methods accordingly. She explained:

> I take a class, gather the data anonymously [in aggregate] for the class in terms of who is verbal, who is visual, what fraction are sensing and intuiting. I think it's always helpful to be thinking about what kind of tools will help this class learn the material.

Why Is It Important to Understand Students' Background and Prior Knowledge?

Faculty noted several reasons why they find it important to understand student backgrounds. They mentioned that a student lacking in relevant background knowledge and coursework faces many obstacles in mastering course content. Instructors may use information about student backgrounds to modify what and how they teach. In some cases, instructors use this information to identify and notify students with weaknesses in their background knowledge. In these cases, the instructor may suggest that the student seek remedial tutoring or put off taking the course until they have completed additional preparatory

(continued)

coursework. Information about students' learning styles, interests in science, and career goals are generally used to tailor instructional methods and select content appropriate for or of interest to the students.

PCK 2: Science Curriculum

The second component of PCK relates to the way that content sequencing within and across courses influences students' acquisition of skills and achievement of learning goals. Here, we limit our discussion to how new faculty members build their syllabi and determine content coverage and sequencing within their own courses.

In the initial interview, seven faculty members indicated that they used a syllabus created by others who had taught or were currently teaching the same course. Three years later, a few faculty members still used the pre-existing syllabi, and kept the basic structure intact while modifying the lectures. Most study participants, however, reported that they had created their own syllabi or heavily modified pre-existing syllabi. Two key themes emerged through these responses:

1. Resources for Building a Syllabus

   * Pre-existing syllabus
   * Textbooks and other references

2. Factors that Impact the Courses Syllabus

   * Relationship of the course to other courses
   * Learning goals for the course
   * Amount of material to cover

1. Resources for Building a Syllabus

When first beginning to teach a course, and particularly in the first year of the study, seven faculty members indicated they based their syllabus on a pre-existing syllabus. In some cases, faculty members did not make substantial modifications to this syllabus before teaching the course for the first time. As Susan explained, "This year, it's going to be very much based on the pre-existing syllabus. Then next year, if I see some areas that I want to change, I'll probably try to incorporate that." Ryan took the same approach, and explained that he would "take notes" on "what I like, don't like . . . what material I can cover faster or slower." A few faculty members made substantial changes to pre-existing syllabi before beginning to teach the course.

Four faculty members used other methods to create their syllabi, while still relying on traditional approaches that begin with the content (whether from an existing syllabus, one or more textbooks, or their knowledge of the content) and divide the content across class sessions. Scott, in creating his syllabus, "basically followed the textbook . . . .[and the] chapters I wanted to cover." Linda "used the textbook as a guideline," and "supplemented" this content "with papers that . . . are relevant to whatever we're talking about . . . and lecture from previous professors." As with the faculty members who began with a pre-existing syllabus, these faculty members described a content-driven syllabus process rather than the learning goal-driven process that backward design recommends (Wiggins & McTighe, 1998).

2. Factors Impacting How Faculty Members Build their Syllabi

The processes by which faculty members built their syllabi differed depending on the class they were assigned to teach and whether the course was one of many sections of a single course taught by different instructors, a part of a sequence of courses, or a prerequisite to other courses in the curriculum. For example, John taught a course with multiple sections offered by different professors, and he noted that he had to use the same textbook, grading system, and syllabus as in the other sections. Robert pointed out a benefit of this approach: "This is a standardized topic that the students need to know before they get to the next level. It sort-of makes sense . . . because you know what has to be taught is standard."

David, who taught the only section of his course, did not have this constraint. He explained that he "inherited" the course from another instructor, but indicated that he was "revamping the whole class" because "our styles are very different and our strengths in different areas are very different." In the follow-up interview, he stressed that he had "been shuffling things around for the past four years" to get "the proper sequence of the concepts."

PCK 3: Instructional Strategies

The third PCK component relates to instructional strategies, such as inquiry-based learning, cooperative and collaborative learning, and technology-aided instruction. In this section, we analyze faculty responses related to the teaching and learning approaches that faculty members planned to use (initial interview) or used (follow-up interview) in their classes.

Research indicates that many faculty members replicate the instructional methods that they experienced as undergraduates (Austin, 2011; Marbach-Ad, Schaefer, & Thompson, 2012; van Driel, Beijaard, & Verloop, 2001). To better understand the different teaching approaches that our new faculty members experienced when they were students, we asked them if they had been exposed as undergraduates to teaching approaches other than lecture. Ten of 11 faculty members indicated that most or all of their undergraduate instruction consisted of "traditional" or "classical" teaching approaches. David explained, "Essentially, professors wrote on the board and we copied." Amber reinforced this point: "I can't think of a course [in which] I had anything other than lecture." Faculty responses related to non-lecture courses were more varied. Amber and Dana noted that their lab courses were "traditional" and "cookbook." Tim had a different experience, however, and described his lab courses as "being scientists . . . you'd have your botany classes where you would actually go out and wander around and look at things first-hand."

Despite their limited experience with nontraditional, student-centered approaches as students, in the initial interview all new faculty members indicated that they intended to implement more varied teaching approaches. However, they often had not fully formulated a plan for implementing such approaches. As Linda explained, "I would love to use different approaches for teaching. I'm not sure if I could come up with them myself." Other faculty members had ideas about teaching approaches that they wanted to use, but were not sure how to implement them. For example, Jenna said, "I've been thinking about using this case study approach. I haven't quite formulated how that will work yet, but it's something I'd like to explore." A few faculty members had more defined plans, which were generally based on their prior teaching experience or on how the class was organized prior to their arrival. Amber inherited a class in which case studies were "already set up," and she planned to continue using this "long case study that lasted three weeks."

In the follow-up interviews, all faculty members reported using some evidence-based, student-centered teaching approaches, and most faculty members had used a variety of different approaches. The pages that follow illustrate the wide variety of non-lecture techniques adopted by these instructors in the first few years of university teaching. These techniques included:

1. In-class discussions
2. Group work
3. Case studies and/or problem-based-learning (PBL)
4. Visual-based instruction (e.g., streamed video or DVD)
5. Critical writings (e.g., reflective journals, summaries, essays, critiques)
6. Games, simulations, and/or role play
7. Directed research
8. Out-of-class discussions (e.g., virtual chat, online discussion board)

1. **In-Class Discussions**

All faculty members reported that they used and valued in-class discussions, but sometimes faced challenges in engaging all students. Amber explained that the type of discussion might vary "depending on the course and depending on the size of the course." Large enrollment courses were more likely to have whole-class discussions than small group discussions, but these discussions did not always work as well as faculty members would have liked. David observed that physical facilities impacted the level of student engagement in the in-class discussion. He shared an example of group discussions among students that occurred "inadvertently" when his class moved from an auditorium-style lecture hall to a room in which the chairs were arranged in circles. He suggested that this "configuration did help quite a bit because they were all facing each other instead of me."

2. **Group Work**

Eight faculty members reported that they asked students to engage in group work inside and/or outside of class. Common group activities included laboratory exercises, projects, oral presentations, poster sessions, and collaboration on homework. Jenna noted that group work activities were not always well-received: "I find that sometimes students get frustrated working in groups, so I [let them start] working independently." The difficulties associated with group work seemed to limit the frequency with which faculty members used this approach, despite national recommendations that highlight the value of group work (Marbach-Ad, Shaefer-Ziemer, Orgler, & Thompson, 2014).

3. Case Studies and/or Problem-Based Learning

Encouragingly, eight faculty members reported using case studies and/or problem-based learning, which are nationally recommended practices. Some faculty members indicated that they used in-depth case studies to tie together various elements of the course. Amber, who inherited a course that already employed a case study approach, explained how she used case studies "to really cement concepts, and even introduce them to new concepts [related to] what's being done in the lecture portion and in the lab." While some faculty used hypothetical situations, others incorporated scientific papers into case studies based on a "real situation" (Tim).

Some faculty members suggested that their courses were particularly well-suited to problem-based learning, while others felt that the value of this approach was independent of the nature of the course. Robert expressed that "my personal philosophy is that students are better taught with problem-based [learning], because as they go through the problems, it brings out what they know about all the concepts."

4. Visual-Based Instruction

Eight faculty members indicated that they used visual-based instruction, generally by incorporating animation or video clips into their lectures or slideshows. Dana explained how she used a "very cool cell animation" to bookend her lecture on this topic: "I show it at the beginning and I try to show it at the end . . . At the beginning they may not recognize anything . . . Hopefully at the end, they can recognize what's going on in the cell." David occasionally used short videos at the end of his lectures. These videos "complement" what was covered in class and "fill in some gaps" with "the little details."

Most faculty members who used visual-based instruction sought available resources on the internet through sites such as YouTube (www.youtube.com) and Nature (www.nature.com). While the availability of these resources has grown dramatically in the past few years, faculty still find it difficult at times to find videos appropriate to their specific needs. As David explained, "I use movies and animation when they are available. I wish they had more, but I need something specific and [sometimes] there's just nothing out there."

5. Critical Writings

Approximately half of the faculty members in the study indicated that they asked students to engage in some sort of critical writing in their courses. This generally involved assigning forms of writing authentic to scientific disciplines, such as analyzing a journal article (Dana, Jenna, and Susan) or writing a research proposal (Linda). Amber described her creative approach to encouraging students to write about scientific topics:

> They read something about synthetic biology and they had to write a letter to Darwin explaining what synthetic biology was.

Some professors indicated that they did not require writing assignments because they were time-consuming to grade. As John lamented, "I wish I could... but it's a matter of time."

6. Games, Simulations, and/or Role Play

Six faculty members reported using games, simulations, and/or role playing. For example, Amber reported using a game to highlight ethical issues in science. She had also used a Jeopardy-like game in her class, and noted that students "love it... but it takes time to make it up." Jenna reported using a role play activity in which students were assigned to represent different enzymes in an enzymatic pathway. A few study participants indicated that they made a deliberate decision not to include games. Susan explained:

> I don't believe in games at this level. They're not here to play; they're here to learn. Maybe that's old school, but I don't want them to think this is high school. They've got to take responsibility for their own learning.

Tim suggested that games could be valuable in lower-level courses that seek to increase student interest in the subject. However, he "made a point not to" incorporate games in the upper-level course that he taught because "I'm emphasizing that the games are over and it's a little bit more real world."

7. Directed Research

Four faculty members indicated that they used directed research in their classes. The directed research activities that faculty mentioned generally overlapped with previously discussed instructional strategies, in that they involved scientific literature and/or group work. Faculty members used directed research to introduce students to and engage them in the process of scientific research. For example, Susan developed a project on the drug design process in which she gave students "background information" and then "they go into the literature and actually look at what has been done by other people, and then apply it to an organism that nothing is known about." David explained that he uses a class project in which students engaged in progressively independent research as "a way of teaching students that they can ask questions, and giving them the tools to do that."

Despite valuing directed research as an instructional approach, some faculty members faced barriers in implementing it in the classroom. Amber, an instructional faculty member who had only teaching responsibilities, commented that the fact that she did not work in a research lab was a barrier to incorporating more directed research into her instruction. As she explained, "I'm not doing primary research so I don't go any further." She hoped to address this limitation by seeking out more collaboration with Tim, a colleague whose research was closely related to the content of Amber's course. However, incorporating this research into her course was still "a matter of finding the time."

8. Out-of-Class Discussions

Four faculty members reported that they had used out-of-class discussions, and they had mixed feelings about its success. Linda found that her students were highly engaged in a course Facebook group and used this forum "all the time ... " As she explained, "I'm amazed how well it works to be on Facebook. They find journals, they find things that they share on their own ... They posted [their video project] on YouTube and then they posted the links on Facebook so everybody could see everybody's video and comment." Not every attempt to incorporate out-of-class discussions shared this success. Susan and Amber had tried online discussion forums but discontinued the practice because of limited student participation.

Summary

As the preceding pages indicate, the cohort of new faculty members greatly increased their repertoire of instructional techniques during their first 3 years at our university. They demonstrated a strong desire to increase the diversity and effectiveness of instructional methods, and many incorporated a number of nationally recommended teaching practices into their courses. However, time continued to present a major limitation to their ability to integrate evidence-based teaching approaches. Time constraints impacted faculty members' ability to determine which methods would be appropriate for their courses, to effectively implement new methods in the classrooms, and to develop the resources that these new methods may require.

PCK 4: Assessment of Student Learning

In order to understand how the new faculty members assessed student learning of subject matter, we asked about the components that comprised student grades in the course and whether faculty members used any modes of assessment other than traditional, high stakes summative assessments comprised of multiple choice, short constructed responses, and/or extended constructed response items.

In the initial interviews, some new faculty members reported that they were not yet sure how they would grade students. As David explained, "I haven't decided. I need to do that in the next four weeks because they will need to know on the first day exactly how I am going to grade them."

Many faculty members, including faculty members who had already decided how they would assess students, indicated that they were not confident in their knowledge of assessment and would like to develop further expertise in this area. As Tim put it, "This is probably where I am the least experienced." One of the specific concerns that faculty members referred to was targeting the appropriate level of difficulty.

David noted, "I am afraid I will give a test that will be too hard," while Amber expressed her view that "it's easy to write an easy multiple choice question, but [to] write a college-level, 400-level course multiple choice question is difficult." Linda, who had taught courses prior to the initial interview, acknowledged that assessing student learning was "something I have trouble with."

In the follow-up interview, all faculty members had devised ways of assessing student learning. Despite growth in their knowledge of assessment, some faculty members commented that this aspect of teaching continued to challenge them. Linda noted, "Assessment is always a problem for me." Other faculty members reported that they still had difficulty with certain aspects of grading, such as determining individual grades for group assignments (Susan) or writing assessment items (David). Faculty members addressed these difficulties in different ways, generally through ongoing revision of their assessment tools and grading conventions, or taking advantage of exams that were developed by other instructors who had previously taught the course. For example, David reported that the "first exam is a very difficult exam to put together... So I reuse the same exam, year after year, because it's really painful for me." Jenna reported that she "leaned pretty heavily on existing people here" in terms of how to assess student learning.

Faculty members described a variety of assessment tools that they used in their classrooms. All of these assessments were graded, although the degree to which they contributed to the students' final grade in the course varied, with some being low-stakes (worth only a small percentage of the entire course grade) and others high-stakes (worth a substantial percentage of the course grade). Below we describe the major forms of assessment used by faculty:

1. Exams
2. Quizzes and Clicker Questions
3. Participation
4. Presentations and Posters
5. Homework Assignments

1. **Exams**

All faculty members used exams as a principal component of grading, and they were in consensus that exams should be a component of measuring student learning. These exams generally constituted the majority, but not the entirety, of student grades. As Ryan explained, the exam is "a big part but it's not the only thing" on which students are graded.

In order to more fully gauge student understanding, faculty members used a variety of item types on their exams. Some faculty members, when using multiple-choice items, asked students to explain their thought process for selecting a response. Faculty members also reported using item types such as true-false, fill in the blank, and short constructed response questions. Tim noted that his exams include "literature-based questions" as well as "a piece of data, a figure, or a series of figures, and then questions related to that." He acknowledged that these questions "took a long time to grade," but he valued them because they assessed students' ability to interpret data and understand the "larger point." Linda similarly highlighted the value of short constructed response items: "They really give me a much better idea about whether students get the concept or not."

Faculty testing strategies differed in terms of how much content they covered on each exam, how frequently exams were given, and whether exam content was cumulative. David explained why he opted to use cumulative assessments in his course: "The content is cumulative. They cannot forget what they studied at the beginning of the semester ... They need to integrate all the concepts, so they're responsible for what they learn on the first day to the last day of the class." Other faculty members reported using a combination of non-cumulative and cumulative exams.

2. Quizzes and Clicker Questions

Seven faculty members reported that they used quizzes or graded clicker questions to assess student understanding of material covered in prior lectures or out-of-class readings. These assessments also served to motivate students to come to class prepared. As David explained, "This year I started little pop quizzes just because I want them to stay up on the material."

Faculty indicated that quizzes and graded clicker questions were valuable, but they could be time-intensive to administer and grade. Linda shared her thoughts on this:

> I've done many things. I've done daily short quizzes ... like two questions from the previous lecture that they just need to remember a concept, so to keep [it in] their memory. And it turns out that in the end they loved it. They really liked being forced to read the previous lectures because it was easier to stick in their heads. The quizzes I'm not doing anymore. It was a managerial nightmare; it was just really tough to do. I've tried clickers. The technology just failed for me ... If clickers were easier to use, I'd definitely do it again.

3. Participation

Four faculty members graded students on participation. This grade generally constituted a relatively small portion of students' grades. David explained his motivation for grading student participation in class: "[Because of university policy] I can't grade them on attendance, but I want them to be present, I want them to read ... I want real participation. So [grading participation] really motivates them to try to provide input." Tim also used participation grades to motivate students to come to class prepared and actively engage in the course, but he did not penalize students who were not actively participating. As he explained, "participation was a gradable aspect but it can help you only, not hurt you."

4. Presentations and Posters

Four faculty members reported that they graded students on presentations, including presentation of posters that students had created as a class assignment. As Amber noted, "Presentations are great. They're easy to grade and I think the students gain a lot."

Presentations and posters were often designed as group projects, which raised issues as to how to grade these assignments fairly. Some faculty members found it difficult to determine individual grades for student participation in group work. Susan explained her challenges in this area and the different approaches that she had taken to overcome these challenges:

> Originally the idea was to grade them as a group, so if somebody wasn't doing their portion in the group it meant that everyone else had to pick up the slack. That did occur... We had some very good students who were paired with some students who did nothing. So to try to make it more equitable... I tried to basically look at the individual ... Groups got a couple of points docked if somebody else didn't do their part, but they didn't get docked 80 % or something like that... I think what we have to do next year is we have to have 90 % of the poster grade on their individual input and maybe 10 % as a group. So they have some motivation to work as a group, but they're not punished for other people.

5. Homework Assignments

Five faculty members reported using homework or problem sets as a graded component of the course. Jenna explained that students are "typically doing homework every week" and that these assignments are valuable because students are "getting lots of feedback." David indicated that he uses "a lot of little homeworks" for students to practice applying course content throughout the semester.

PCK 5: Orientation to Teaching Science

In order to understand how the new faculty members perceive their teaching role and how they approach science teaching as a part of their broader professional responsibilities, we analyzed their responses related to their affective disposition towards teaching, and the balance between teaching and their other professional responsibilities.

Affective Disposition Towards Teaching

In both the initial and the follow-up interviews, all respondents indicated that they had positive feelings about teaching. Many of them were very enthusiastic, and used phrases such as "I love teaching" (Amber, Dana, John, Linda), "it's really exciting" (Jenna), and "I enjoyed interacting" with students (Scott, Amber). Respondents highlighted different aspects of teaching that they enjoyed or found valuable. Jenna, for example, explained that she enjoyed teaching because "it's really exciting to teach people and explore new ideas and, again, I do always learn things." Robert commented that he enjoyed the broad perspective that teaching gave him: "I see teaching as the way to get out of my own research. If you only do your research, you are only looking at one area, but if you teach you teach a broader [range of topics] ... It makes you look at the literature." Some faculty members explained that they enjoy teaching because of the impact it has on students. As Susan phrased it, "I really love when I get to do a lecture or a discussion or something where I'm really interacting with the students because I love science . . . I really like to pass on my enthusiasm for science to other people and say, 'This really is cool; this really is interesting.'"

While no faculty members expressed predominantly negative feelings about teaching, several indicated that there were some aspects of teaching that they disliked. These mixed feelings generally arose from discomfort with particular aspects of instruction. John explained that he experienced anxiety while teaching: "I'm very concerned that the students will ask something that I don't know and I get anxious about it, so I try to make sure that I know everything." Jenna described herself as a "bashful person," and she noted that "it [takes] a lot of energy for me to get in front of a group, so it's not an easy thing for me to do."

A few faculty members reported that they enjoyed the instructional aspects of teaching, but not some of the non-instructional aspects that accompany it. Dana commented that she found "procedural issues" related to student absences to be "very frustrating." She indicated that "it would be nice if there were more uniformity" and faculty members "had the same policies on absences— unexcused, excused, illness—all that stuff." Other faculty members noted that they sometimes felt poorly equipped to support students in non-academic issues. As David explained, teaching requires faculty members to address a range of challenging student concerns, such as mental health issues. In the context of a growing concern for the mental health and coping skills of young adults, and faculty members' regular interactions with large numbers of students, faculty members may be among the first to notice such issues. However, many faculty members feel unprepared to address these types of non-academic issues, and they generally have limited knowledge of the resources available to help students through these kinds of challenges. As David noted, "These are things that you don't get trained to do. And they're very difficult to do."

Concerns About Balancing Responsibilities

In the initial interview, all new faculty members expressed some degree of concern about balancing their research, teaching, and other responsibilities, with the exception of two new faculty members who did not have any research responsibilities. Linda, who was new to teaching, acknowledged that she "has no idea" how to balance these tasks and that she would "see what happens." Other faculty members had some teaching experience and better understood what their teaching responsibilities would entail, but still expressed concern about striking a balance between competing responsibilities.

In the follow-up interviews, eight of the nine faculty members with both teaching and research responsibilities noted that balancing the two was difficult, particularly when coupled with other responsibilities. In the words of Tim, "There are times when the expectation can overlap . . . and it can really be tough in terms of service and research . . . When you're teaching . . . you'd love to just do it and have no other distractions."

Faculty members highlighted different responsibilities that compete with teaching responsibilities for their time. Multiple faculty members noted that "it's very difficult to balance" teaching and grant-work. John explained that his balance is "okay" because he was not at that time under pressure to seek out additional grant money. However, he was concerned that this balance might not hold. Scott commented that "the difficulty that I find is not so much between the teaching and the research, but it's between all the other things that crowd in ... The difficulty is that there's all sorts of committee work, and emails to answer, and this and that paperwork."

Faculty Approaches to Balancing Multiple Responsibilities

In both the initial and follow-up interviews, many faculty members noted that they prioritize teaching when the two responsibilities are in conflict. Some new faculty indicated that they chose this approach because teaching was a new task for them. As Scott explained, "Because I'm teaching for the first time ... that inevitably becomes the priority." However, even after he had taught for a few years, he indicated that "when I'm teaching, that has to be my first priority" because being in front of a class a couple of times a week "forces me to be prepared." Linda took a similar approach, and in the initial interview she explained that work she had done in the previous year would facilitate this balance:

> Because I had a year off [from teaching], I spent a long time in doing research and I wrote
> a couple of grants already . . . I'll probably give a lot of priority to teaching because it's my
> first time and I want to make sure I do the students justice to their time and effort.

As with Scott, Linda continued to focus a lot of time on her teaching responsibilities
3 years later. She acknowledged that, as an assistant professor who was on the tenure
track, she had "a huge burden to get research done," but she was "following what I
think is right and I'm giving a lot more time to my teaching than I should."

A couple of faculty members anticipated that balancing their various responsibilities would require them to work long hours. David exemplified this view:

> The teaching will need a time commitment, my research and my grant applications will
> need a time commitment. Unfortunately, I am going to have to resort to the methods I've
> always used, which is to make my day a 14-, 15-hour day, stop it, go, do my family thing
> and then when everybody goes to bed at 8:30 pm, go back and do another four hours.

In the follow-up interview, David reported that his "workday is well beyond the
eight hours." Susan mirrored his sentiment, and explained that she prepared for class
outside of traditional work hours. She noted that she taught "three times a week."
This in-class time fit into her normal schedule, but preparing new lectures did not.
As she explained, "This is the fourth year I'm going to have to do [a course] from
scratch. Unfortunately, I have to [put the lectures together] on my own personal
time at home, so I can spend my time in lab during the day and with my graduate
students."

Through prioritizing or putting in extra time, faculty members generally indicated that they were able to dedicate sufficient time to their teaching responsibilities
despite their initial concerns about time constraints and balancing multiple responsibilities. As Tim acknowledged, he was "always concerned" and felt like he was
sometimes "under the gun" to fulfill a teaching responsibility. However, despite this
pressure, he noted that "there's always [a way] I can do what I need to do and still
be accessible both to the students [in my class] and people in the lab."

Professional Development for New Faculty Members

In this section, we discuss the implications of these study findings for new faculty
professional development. The initial interviews demonstrated that faculty members
start their new positions with varied amounts and types of experience. All faculty
members, regardless of prior experience, arrived with some concerns related to
their teaching responsibilities. During the 3-year period, faculty members sought
and benefitted from multiple types of professional development, including formal
activities as well as structured and unstructured learning networks.

Based on their experiences over their first 3 years, faculty members suggested
that professional development programs for incoming faculty should include multiple components (Marbach-Ad et al., 2013):

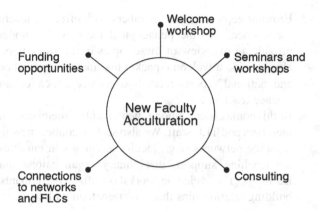

Fig. 3.1 The TLC's acculturation activities for new faculty members

- Training in teaching during the earliest career phases (e.g., graduate school and postdoctoral appointments)
- Examples and models of good practices in teaching science
- Topical seminars (e.g., concept maps, using technology in the classroom, assessing student learning)
- Mechanisms to gauge students' background knowledge
- Mentoring from experienced faculty members
- Assistance with the course (e.g., undergraduate learning assistants, graduate teaching assistants, postdoctoral fellows)
- Feedback on teaching from experienced faculty members

A teaching and learning center can provide or support all of these types of professional development. A center with a disciplinary focus is particularly well suited to this task because the staff possess in-depth knowledge about activities, networks, and resources that integrate content and pedagogy.

Our teaching and learning center offers a menu of opportunities (Fig. 3.1) that aim to acculturate faculty to their new institutional context. These include some professional development specifically designed for this population, as well as activities that meet their needs while serving the broader instructional community.

Welcome Workshop

Every year, the TLC conducts a welcome workshop for new faculty members. To prepare new faculty members for success in their role as science educators, we designed the workshop to achieve three primary goals:

1. Introduce incoming faculty members to the most relevant literature on science education at the undergraduate level. Within this literature, we share information about national concerns related to science education and the role of undergraduate instruction in improving science education.

2. Familiarize new faculty members with effective teaching and learning methods for science, and the pedagogical theories that undergird those methods. We provide an overview of these topics in the workshop. For deeper coverage, we also provide a welcome packet that includes reference books, scholarly papers, and national reports related to science education and effective practices for science teaching.
3. Build connections among new faculty members, and between new faculty members and TLC staff. We also seek to connect new faculty members with other resource networks (e.g., faculty learning communities, campus-wide resources for teaching support, disciplinary organizations that promote teaching and learning). The welcome workshop offers participants an opportunity to begin building relationships that can strengthen with time.

This workshop has strong support from departmental and College leadership. College and departmental leaders demonstrate their support in multiple ways, including promoting new faculty member attendance at the workshop, providing funding for workshop materials and food, and attending the workshop to reinforce the TLC's message about the importance of effective teaching.

More information about the welcome workshop, including an outline of workshop topics and welcome packet contents, can be found in the Implementation Guide.

Seminars and Workshops

We encourage new faculty to participate in our regularly offered seminars and workshops on science teaching and learning (see Chap. 2 for a detailed description of TLC seminars and workshops). We also publicize relevant workshops hosted by other campus entities, such as the Center for Teaching Excellence, which offers workshops to assist faculty in developing pedagogical competence, and the Office of Learning Technologies, which offers training and technical support related to incorporating technology in instruction. Based on feedback from the new faculty interviews, these seminars and workshops have helped new faculty members to develop specific skills and also provided an opportunity for them to build connections with like-minded colleagues. We also found some topics, such as clickers and other instructional technology, to be particularly timely and valuable for our faculty members.

Individual Consulting

New faculty members who require additional or specialized support may schedule an appointment with TLC staff for individual consulting (see Chap. 4 for a detailed explanation of the TLC's individual consulting services). New faculty members may contact the TLC at the suggestion of experienced faculty members and department chairs. New faculty members generally seek individual consulting related to a specific teaching issue, such as implementing a new pedagogical technique in their class or adjusting their teaching in response to negative feedback from student evaluations. Through consulting sessions, we aim to provide faculty members with the skills and resources they need to develop competence as instructors.

Connections to Networks and Faculty Learning Communities

New instructors are introduced to and encouraged to join on- and off-campus faculty learning communities (FLCs; see Chap. 4 for more details) that are relevant to their teaching interests. For those faculty with interests that are not well-served by existing communities, we have on occasion helped them establish a new community. Of the 11 faculty members in the longitudinal study, seven participated in at least one FLC. These new faculty members indicated that their participation in a community was very helpful. They appreciated the regular, ongoing contact with their colleagues and the time dedicated to focusing on their teaching that the FLC provided. They also indicated that FLCs provided them with opportunities to build relationships and be mentored by their more experienced colleagues.

Funding Opportunities

We support new faculty members both through internal funding and assisting them in securing external funding. Our internal funding is targeted at new faculty members who want to attend off-campus seminars, workshops, and conferences on teaching and learning. Our assistance for securing external funding involves helping new faculty members identify appropriate funding sources (e.g., university course redesign grants, NSF CAREER grants), assisting them in developing proposals, particularly the educational aspects of the grant; and providing letters of support for grant proposals.

Conclusion

While the primary goal of the longitudinal study was to inform our professional development offerings for new faculty members, the rich qualitative data that the study provided also have implications for professional development offerings for experienced and future faculty members. Study findings informed our understanding of the need for ongoing professional development for experienced faculty members. In Chap. 4, we describe how the TLC's consultation services fill this role. Our research also highlighted areas where we can enhance our training program for preparing graduate students, some of whom are the faculty of tomorrow, for their teaching responsibilities. To this end, we now offer both mandatory preparatory courses for all new biology and chemistry graduate teaching assistants, as well as optional long-term programs for graduate students who seek to further develop their expertise in teaching and learning. These programs for graduate students are detailed in Chap. 5.

Implementation Guide

On the pages that follow, we provide materials to support the implementation of professional development activities for new faculty members.

Longitudinal Study: Interview Protocols

Our longitudinal study consisted of pre- and post-interviews with a cohort of new faculty members. Interview protocols are below.

Initial Interview Protocol

1. Did you have any prior experience as a teacher (TA, guest lecturer, etc.)?
2. What is your teaching philosophy?
3. What should chemical and life science majors acquire in their undergraduate studies?

    Note for the interviewer—probe for the following aspects after initial response:

    (a) Content knowledge
    (b) Laboratory skills
    (c) Scientific writing and reading
    (d) Understanding the dynamic nature of science
    (e) Understanding the applicability of science to everyday life
    (f) Understanding what scientists do
    (g) Historical perspectives
    (h) Motivation to continue in the field

4. As a student (undergraduate/graduate), were you exposed to approaches for teaching and learning other than lecture? Do you think that you will use them in your teaching? If so, how?

    Note for the interviewer—probe for the following aspects after initial response:

    (a) Case studies and/or problem-based learning
    (b) In-class discussions

(c) Out-of-class discussions (e.g., virtual chat, online discussion board)
(d) Critical writings (e.g., reflective journals, summaries, essays, critiques)
(e) Group work
(f) Visual-based instruction (e.g., streamed video or DVD)
(g) Games and simulations
(h) Problem-solving
(i) Online presentations
(j) Role play
(k) Directed research

5. As an undergraduate student, how did you interact with your instructors after classes? (e.g., email, office hours, web, question box)
6. How do you think that you are going to interact with your students?
7. How do you envision your class sessions? (e.g., review of the last session in the first 10 minutes, time for questions)
8. How do you build your syllabus? Do you use the online learning management system?
9. How do you build the course assessments? Do you use any assessments other than traditional exams?
10. How do you learn about your students' background knowledge in order to relate to the diversity in the class?
11. How do you feel about teaching? Do you like teaching? Why or why not?
12. What do you think would help you to prepare to teach your first class? (e.g., workshops on teaching and learning, mentoring from your colleagues, participating in a support group)
13. How do you plan to divide your time between research and teaching? Are you concerned about being able to do a good job at both?

Follow-up Interview Protocol

1. How many semesters have you taught at the University of Maryland?
2. What type of classes did you teach? (e.g., majors, non-majors, introductory, upper-level, small, large enrollment)
3. What is your teaching philosophy?
4. What should chemical and life science majors acquire in their undergraduate studies? What are you doing in order to make sure that you are giving the students the opportunity to acquire these?

   Note for the interviewer—probe for the following aspects after initial response:

   (a) Content knowledge
   (b) Laboratory skills
   (c) Scientific writing and reading
   (d) Understanding the dynamic nature of science
   (e) Understanding the applicability of science to everyday life

(f) Understanding what scientists do
(g) Historical perspectives
(h) Motivation to continue in the field

5. Are you using the following in your teaching? If so, how?

(a) Case studies and/or problem-based learning
(b) In-class discussions
(c) Out-of-class discussions (e.g., virtual chat, online discussion board)
(d) Critical writings (e.g., reflective journals, summaries, essays, critiques)
(e) Group work
(f) Visual-based instruction (e.g., streamed video or DVD)
(g) Games and simulations
(h) Problem-solving
(i) Online presentations
(j) Role play
(k) Directed research

6. How do you interact with your students?
7. How do you plan your class sessions? (e.g., review of the last session in the first 10 minutes, time for questions)
8. How do you build your syllabus? Do you use the online learning management system?
9. How do you build the course assessments? Do you use any alternative assessments?
10. How do you learn about your students' background knowledge in order to relate to the diversity in the class?
11. How do you feel about teaching? Do you like teaching? Why or why not?
12. Did you use any resources inside or outside of the university to learn skills to enhance your teaching skills? Explain.
13. Are you using an interdisciplinary teaching approach?
14. Do you belong to any community organized around teaching, either at the university or outside of the university?
15. How do you divide your time between research and teaching? Are you concerned about being able to do a good job at both?

Welcome Workshop

The new faculty welcome workshop program is detailed below. We have found that this sequence of topics works well for new faculty at our institution, but the sequence and combinations of topics can be customized to meet the specific needs of other institutions. Since STEM teaching and learning is now an area of prolific research, we also recommend periodic updates to encompass more recent scholarly literature.

We find it valuable to begin with a discussion of the problem of STEM student success and persistence (the so-called "leaky pipeline"), because this frames the need for improving undergraduate science education and presents a broad perspective on the role of science educators. Here, we draw heavily on Seymour and Hewitt's (1997) landmark analysis and the more recent PCAST report (2012).

With this foundation, we then move to a ground-level approach. We present learning theories, evidence-based teaching approaches, and teaching tips. We conclude the presentation by pointing new faculty members to additional resources. Throughout the presentation, we reiterate that our Teaching and Learning Center is available to assist them with any teaching issue they encounter.

Welcome Workshop Outline

1. The Leaky Pipeline

Many undergraduates who plan to major in science, mathematics, or engineering do not graduate with a degree in these fields. Seymour and Hewitt (1997) found that approximately 20 % of students who planned to major in mathematics and the physical sciences leave the field. The percentage of students completing degrees in other STEM fields is only marginally higher, at 40 % for engineering and 50 % for the life sciences. A more recent report indicates that this problem persists, with fewer than 40 % of students who enter college with plans to graduate in a STEM field obtaining a STEM degree (PCAST, 2012).

Seymour and Hewitt (1997) conducted interviews with both "switchers" and "persisters" from seven different institutions to explore why some students leave STEM fields while others persist. They found that students who switched out of STEM majors were not less talented academically than their peers who persisted; instead, the groups were very similar in terms of individual attributes of performance, behavior, and attitude. Both persisters and switchers shared impressions of an inhospitable STEM learning environment (see Box: Reasons for Leaving the Sciences for more details). Within this context, we stress to new faculty members the importance of being aware of these potentially problematic characteristics of the STEM learning environment, because they can play key roles in institutional change efforts that aim to rectify these pipeline issues.

Reasons for Leaving the Sciences (Seymour & Hewitt, 1997)

The 'hardness' of science: Students noted challenges related to conceptual difficulties, the amount of material they were expected to master, and the pace at which the material was presented.

The significance of grades: Receiving poor grades in classes in the STEM major, and particularly in introductory classes, contributed to attrition in these fields.

(continued)

The competitive culture: National norms of science education favor individual competition over collaborative learning. Some students found this culture to be uncomfortable or offensive. Students also suggested that the competitive culture promotes striving for high grades rather than deep learning.

The weed-out tradition: All institutions exhibited an implicit or explicit weed-out culture with expected attrition rates of 40–50 % for students in introductory courses. Some students expressed that the faculty members who taught these courses were unsupportive of their learning, in part because they expected a large percentage of students to fail.

The unsupportive culture: While students sought help on a variety of topics, many reported that they failed to receive the needed advice, counseling, or tutorial help.

Problems with faculty pedagogy: Poor teaching was by far the most common complaint. Students who had this complaint strongly believed that STEM faculty members do not like to teach, do not value teaching as a professional activity, and lack incentives to teach effectively.

2. Learning Theory and Evidence-based Teaching Approaches

New instructors face a number of challenges as they begin teaching. These challenges often emerge in the process of determining what to teach, how to teach, how to address the diversity of their students, and how to assess student learning. Below, we provide theory and suggestions to assist new faculty as they tackle these tasks.

(a) What to teach (content and curriculum)

When developing curriculum, new faculty should consult with colleagues who teach prerequisite, parallel, and/or subsequent courses in the curriculum. We also introduce new faculty members to learning theories that can help them make decisions about how to structure their curriculum (see Box: Key Learning Theories for topics to include).

Key Learning Theories
Constructivism: Knowledge is built through manipulation and transformation of what we already know (Ausubel, 1968; Bruner, 1960; Dewey, 1897; Piaget, 1954).
Alternative conceptions: Students often inappropriately generalize their everyday knowledge and knowledge from prior coursework (Fisher, 1983; Gilbert, Osborne, & Fensham, 1982; Thijs & van den Berg, 1993).

(continued)

Spiral curriculum: A spiral curriculum involves periodically revisiting each subject or skill area, with the level of sophistication increasing over time (Bruner, 1960).

Learning progressions: Learning progressions describe how students develop knowledge over time and move from novice to expert understanding (Duschl, Maeng, & Sezen, 2011; Smith, Wiser, Anderson, & Krajcik, 2006).

(b) How to teach (pedagogy)

There is a growing consensus around the superiority of student-centered, active learning approaches over traditional teacher-centered approaches. Varied approaches can be used to promote active learning in science classes, and educators should use approaches that are appropriate to the level, content, and physical setting of their course. These approaches include case-based learning, model-based learning, group work, role playing, games and simulations, reading and writing activities, and authentic research. For more information about these approaches and resources for implementing them in the classroom, see AAAS Vision and Change (2011), Handelsman et al. (2004), and Svinicki & McKeachie (2010).

(c) How to address student diversity (learning styles)

Students come into a course with varying levels of background knowledge, experience, and interest in the subject. Some researchers (Felder, 1993) suggest that students also vary in terms of how they process knowledge and best learn new material. We suggest that faculty members probe student backgrounds in terms of prior coursework, major, specific knowledge and skills related to the content of the course, and/or their interests and learning styles (see Box: Felder's Learning Styles). Understanding student diversity allows faculty to tailor the course curriculum and level of difficulty to student knowledge, skills, interests, and learning styles. Faculty members should also be sensitive to differences related to student race, gender, cultural background, and personality (Svinicki & McKeachie, 2010).

Felder's Learning Styles

Students have a variety of preferences for how they learn best. Felder (1993) suggests that students' learning styles can be defined through the following dimensions, each of which are measured along a continuum:

(a) Sensing and intuitive perception
(b) Visual and verbal input
(c) Inductive and deductive organization

(continued)

(d) Active and reflective processing
(e) Sequential and global understanding

Felder's Index of Learning Styles Questionnaire can be taken online and provides immediate feedback on preferred learning style (www.engr.ncsu. edu/learningstyles/ilsweb.html). Although there is not universal acceptance of the learning style theory, Felder's dimensions can be a useful construct for planning a varied curriculum.

(d) How to assess student learning (assessment)

A growing body of research recognizes the benefits of using multiple methods to assess student learning (Bransford, Brown, & Cocking, 2000; Cole, 1991; Resnick & Resnick, 1992; Walvoord & Anderson, 2010). Ideally, these methods include both formative assessments, which provide students and instructor with immediate feedback on their progress, as well as summative assessments, which generally measure mastery of knowledge and skills at the end of a teaching unit and/or course. We provide new faculty members with an overview of different types of assessments, such as traditional multiple-choice tests, attitude surveys, weekly reports, performance assessments, and concept maps. We also refer them to additional resources in the Field-tested Learning Assessment Guide (FLAG; www. flaguide.org/cat/cat.php), developed by the National Institute for Science Education. The guide offers an overview of various assessment tools, includes examples of items for each type of assessment, provides suggestions for implementation, and discusses the advantages and disadvantages of the various techniques. We also suggest that faculty members become familiar with Bloom's Taxonomy of Learning (Bloom, 1984) and more recent theory built on Bloom's Taxonomy (Mayer, 2002). These theories suggest that assessment should measure multiple cognitive levels and reflect the progression of cognitive skills necessary for the mastery of course content.

3. Resources Available to New Faculty Members

We refer new faculty members to a variety of resources for additional support as they adjust to their new instructional responsibilities. Our goal is to familiarize them with resources that cater to their varied interests, time availability, and preferred mode of professional development. These include:

- **Campus workshops on teaching and learning**
- **TLC seminars and workshops** (see Chap. 2)
- **Experienced faculty members who can serve as teaching mentors**
- **Faculty Learning Communities** (see Chap. 4)
- **Science education literature and professional societies**

Science Education Literature and Networks
New faculty members who are active researchers routinely immerse them-selves in the literature of their scientific discipline but are largely unaware of the literature of science education. The TLC houses a library of science education books and journals, compiles online resources on its website, and encourages faculty to come to the TLC with specific requests and questions.

Annual conferences: The TLC maintains a frequently updated list of upcoming conferences and meetings related to science education (www.cmns-tlc. umd.edu/conferences). This list includes general information about each conference, such as focus, logistics, and abstract submission instructions.

Databases: We introduce new faculty members to science education databases such as the Education Resources Information Center (ERIC; www. eric.ed.gov) and the National Science Digital Library's BioSciEdNet (www. biosciednet.org).

Journals: The TLC website includes a list of journals that cover topics related to science education (www.cmns-tlc.umd.edu/journals). TLC staff provide faculty members interested in submitting articles with specific guidance on each journal's scope and requirements (e.g., allowable length of manuscript, research methods) as well as general guidance on how to write up a science education study.

Books: The TLC maintains a list of helpful books for new instructors (www. cmns-tlc.umd.edu/books).

Welcome Packet

Our welcome packets contain a variety of reports, books, and papers related to STEM teaching and learning. The contents of the packets are periodically revised to reflect influential new publications, the specific disciplinary focus of new faculty members, and current departmental or College change initiatives. At the time of publication, our welcome packet included the following materials:

1. **Influential Reports from National Science Organizations**
   - *AAAS Vision and Change* (2011): This report, produced by the American Association for the Advancement of Science, is intended to serve as a framework for reform of undergraduate biology education. It highlights core concepts and competencies that students should develop over the course of their undergraduate degree programs and recommends evidence-based, student-centered pedagogical approaches. An electronic version of the report is available from www.visionandchange.org.

- *Scientific Foundations for Future Physicians (AAMC-HHMI Committee, 2009)*: This joint report of the Association of American Medical Colleges and the Howard Hughes Medical Institute outlines scientific competencies that students should have achieved prior to enrolling in and after completing medical school. An electronic version of the report is available from www.aamc.org.
- *Making the Right Moves. A Practical Guide to Scientific Management for Postdocs and New Faculty (Burroughs Welcome Fund (BWF) and Howard Hughes Medical Institution (HHMI), 2006)*: This book, produced by HHMI, focuses primarily on new faculty members' responsibilities as researchers, but Chapter 13, entitled Teaching and Course Design, gives a broad overview of effective teaching. An electronic version of Chapter 13 is available from www.hhmi.org/resources/labmanagement/downloads/moves2_ch13.pdf.

2. **Books on Effective Teaching**

- *Scientific Teaching* (Handelsman et al., 2007): This book provides an overview of the theory behind scientific teaching and offers suggestions for bringing this theory into practice in the undergraduate science classroom. The book also includes resource materials for providing professional development that supports science faculty members in improving their teaching.
- *Teaching Tips: Strategies Research, and Theory for College and University Teachers* (Svinicki & McKeachie, 2010): This book, aimed at faculty members in all disciplines, offers a comprehensive and practical guide for teaching. The book covers many aspects of teaching, from techniques for new educators to more advanced methods that enhance experienced educators' repertoire of skills. The book includes practical tips as well as theory and research on teaching and learning.

3. **Papers, Articles, and Other References**

- Learning and Teaching Styles in College Science Education (Felder, 1993)
- Scientific Teaching (Handelsman et al., 2004)
- Clickers in the Large Classroom: Current Research and Best-Practice Tips (Caldwell, 2007)
- Why Not Try a Scientific Approach To Scientific Education (Wieman, 2007)
- Innovations in Teaching Undergraduate Biology and Why We Need Them (Wood, 2009)
- Increased Course Structure Improves Performance in Introductory Biology (Freeman, Haak, & Wenderoth, 2011)
- Understanding Academic Performance in Organic Chemistry (Szu et al., 2011)
- Additional articles that document undergraduate science education research in our departments (Injaian, Smith, German Shipley, Marbach-Ad, & Fredericksen, 2011; Quimby, McIver, Marbach-Ad, & Smith, 2011; Senkevitch, Marbach-Ad, Smith, & Song, 2011)

References

AAMC-HHMI Committee. (2009). *Scientific foundations for future physicians.* Washington, DC: Association of American Medical Colleges. https://members.aamc.org/eweb/upload/Scientific%20Foundations%20for%20Future%20Physicians%20%20Report2%202009.pdf

American Association for the Advancement of Science (AAAS). (2011). *Vision and change: A call to action.* Washington, DC: AAAS.

Austin, A. E. (2002). Preparing the next generation of faculty: Graduate school as socialization to the academic career. *The Journal of Higher Education, 73*(1), 94–122.

Austin, A. E. (2011). *Promoting evidence-based change in undergraduate science education.* A paper commissioned by the National Academies National Research Council Board on Science Education. http://dev.tidemarkinstitute.org/sites/default/files/documents/Use%20of%20Evidence%20in%20Changinge%20Undergraduate%20Science%20Education%20%28Austin%29.pdf

Austin, A. E., & McDaniels, M. (2006). Preparing the professoriate of the future: Graduate student socialization for faculty roles. In J. C. Smart (Ed.), *Handbook of theory and research.* New York, NY: Springer.

Austin, A. E., Sorcinelli, M. D., & McDaniels, M. (2007). Understanding new faculty: Background, aspirations, challenges, and growth. In R. Perry & J. Smart (Eds.), *The scholarship of teaching and learning in higher education: An evidence-based perspective* (pp. 39–89). Dordrecht, The Netherlands: Springer.

Ausubel, D. (1968). *Educational psychology: A cognitive view.* New York, NY: Rinehart & Winston.

Bloom, B. S. (1984). *Taxonomy of educational objectives: Handbook 1: Cognitive domain.* New York, NY: Longman Inc.

Boice, R. (2000). *Advice for new faculty members: Nihil Nimus.* Needham Heights, MA: Allyn & Bacon.

Boice, R. (2011, June.). *Improving teaching and writing by mastering basic imagination skills.* Paper presented at the Lilly conference on College & University Teaching, Washington, DC.

Bouwma-Gearhart, J. L., & Schmid, S. E. (2012, April). *Mixed-methods study investigating research university STEM faculty motivation to engage in teaching professional development.* Paper presented at the annual meeting of the American Educational Research Association, Vancouver, BC, Canada.

Boyer Commission on Undergraduates in the Research University. (1998). *Reinventing under-graduate education: A blueprint for America's research universities.* Stony Brook, NY: State University of New York at Stony Brook.

Bransford, J. D., Brown, A. L., & Cocking, R. R. (2000). *The design of learning environments: Assessment-centered environments. How people learn: Brain, mind, experience, and school.* Washington, DC: National Academy Press.

Bruner, J. (1960). *The process of education.* Cambridge, MA: Harvard University Press.

BWF, & HHMI. (2006). *Making the right moves. A practical guide to scientific management for postdocs and new faculty* (2nd ed.). Research Triangle Park, NC: Burroughs Welcome Fund and Howard Hughes Medical Institution.

Caldwell, J. E. (2007). Clickers in the large classroom: Current research and best-practice tips. *CBE Life Sciences Education, 6*(1), 9–20. doi:10.1187/cbe.06-12-0205.

Cole, N. (1991). The impact of science assessment on classroom practice. In G. Kulm & S. Malcom (Eds.), *Science assessment in the service of reform* (pp. 97–106). Washington, DC: American Association for the Advancement of Science.

Cox, M. D. (1995). The development of new and junior faculty. In W. A. Wright (Ed.), *Teaching improvement practices: Successful strategies for higher education.* Bolton, MA: Anker.

Dewey, J. (1897). My pedagogical creed. *School Journal, 54*, 77–80.

Duschl, R., Maeng, S., & Sezen, A. (2011). Learning progressions and teaching sequences: A review and analysis. *Studies in Science Education, 47*(2), 123–182.

Felder, R. M. (1993). Learning and teaching styles in college science education. *Journal of College Science Teaching, 23*(5), 286–290.

Fisher, K. M. (1983). *Amino acids and translation: A misconception in biology.* Paper presented at the international seminar on misconceptions in science and mathematics, Cornell University, Ithaca, NY.

Freeman, S., Haak, D., & Wenderoth, M. P. (2011). Increased course structure improves performance in introductory biology. *CBE—Life Sciences Education, 10*, 175–186.

Gilbert, J. K., Osborne, R. J., & Fensham, P. J. (1982). Children's science and its consequences for teaching. *Science Education, 66*(4), 623–633.

Golde, C. M., & Dore, T. M. (2001). At cross purposes: What the experiences of doctoral students reveal about doctoral education. Philadelphia, PA: Pew Charitable Trusts. Retrieved from www.phd-survey.org

Handelsman, J., Ebert-May, D., Beichner, R., Bruns, P., Chang, A., DeHaan, R., ... Wood, W. B. (2004). Scientific teaching. *Science, 304*(5670), 521–522.

Handelsman, J., Miller, S., & Pfund, C. (2007). *Scientific teaching.* New York, NY: W.H. Freeman & Company in collaboration with Roberts & Company Publishers.

Injaian, L., Smith, A. C., German Shipley, J., Marbach-Ad, G., & Fredericksen, B. (2011). Antiviral drug research proposal activity. *Journal of Microbiology & Biology Education, 12*, 18–28.

Luft, J. A., Kurdziel, J. P., Roehrig, G. H., & Turner, J. (2004). Growing a garden without water: Graduate teaching assistants in introductory science laboratories at a doctoral/research university. *Journal of Research in Science Teaching, 41*(3), 211–233.

Marbach-Ad, G., Schaefer, K. L., & Thompson, K. V. (2012). Faculty teaching philosophies, reported practices, and concerns inform the design of professional development activities of a disciplinary teaching and learning center. *Journal on Centers for Teaching and Learning, 4*, 119–137.

Marbach-Ad, G., Schaefer Ziemer, K. L., Thompson, K. V., & Orgler, M. (2013). New instructor teaching experience in a research-intensive university. *Journal on Centers for Teaching and Learning, 5*, 49–90.

Marbach-Ad, G., Shaefer-Ziemer, K., Orgler, M., & Thompson, K. (2014). Science teaching beliefs and reported approaches within a research university: Perspectives from faculty, graduate students, and undergraduates. *International Journal of Teaching and Learning in Higher Education, 26*(2).

Mayer, R. E. (2002). Rote versus meaningful learning. *Theory Into Practice, 41*, 226–232.

Nyquist, J. D., Manning, L., Wulff, D. H., Austin, A. E., Sprague, J., Fraser, P. K., ... Woodford, B. (1999). On the road to becoming a professor: The graduate student experience. *Change, 31*(3), 18–27.

Piaget, J. (1954). *The construction of reality in the child.* New York, NY: Routledge.

President's Council of Advisors on Science and Technology (PCAST). (2012). *Engage to excel: Producing one million additional college graduates with degrees in science, technology, engineering, and mathematics.* Available at www.whitehouse.gov/sites/default/files/microsites/ostp/pcast-engage-to-excel-final_2-25-12.pdf

Quimby, B. B., McIver, K. S., Marbach-Ad, G., & Smith, A. C. (2011). Investigating how microbes respond to their environment: Bringing current research into pathogenic microbiology course. *Journal of Microbiology & Biology Education, 12*, 176–184.

Resnick, L. B., & Resnick, D. P. (1992). Assessing the thinking curriculum: New tools for educational reform. In B. R. Gifford & M. C. O'Connor (Eds.), *Changing assessments* (pp. 37-76). Boston, MA: Kluwer.

Reybold, L. E. (2003). Pathways to the professorate: The development of faculty identity in education. *Innovative Higher Education, 27*, 235–252.

Senkevitch, E., Marbach-Ad, G., Smith, A. C., & Song, S. (2011). Using primary literature to engage student learning in scientific research and writing. *Journal of Microbiology and Biology Education, 12*, 144–151.

Seymour, E., & Hewitt, N. M. (1997). *Talking about leaving: Why undergraduates leave the sciences.* Boulder, CO: Westview Press.

Smith, C. L., Wiser, M., Anderson, C. W., & Krajcik, J. (2006). Implications of research on children's learning for standards and assessment: A proposed learning progression for matter and the atomic-molecular theory. *Measurement: Interdisciplinary Research and Perspectives, 4,* 1–98.

Svinicki, M., & McKeachie, W. J. (2010). *McKeachie's teaching tips: Strategies, research, and theory for college and university teachers.* Belmont, CA: Wadsworth Cengage Learning.

Szu, E., Nandagopal, K., Shavelson, R. J., Lopez, E. J., Penn, J. H., Scharberg, M., & Hill, G. W. (2011). Understanding academic performance in organic chemistry. *Journal of Chemical Education, 88*(9), 1238–1242.

Tanner, K., & Allen, D. (2002). Approaches to cell biology teaching: A primer on standards. *Cell Biology Education, 1*(4), 95–100. doi:10.1187/cbe.02-09-0046.

Thijs, G. D., & van den Berg, E. (1993, August 1–4). *Cultural factors in the origin and remediation of alternative conceptions.* Paper presented at the third international seminar on misconceptions and educational strategies in science and mathematics, Cornell University, Ithaca, NY.

van Driel, J. H., Beijaard, D., & Verloop, N. (2001). Professional development and reform in science education: The role of teachers' practical knowledge. *Journal of Research in Science Teaching, 38*(2), 137–158.

Walvoord, B. E., & Anderson, V. J. (2010). *Effective grading: A tool for learning and assessment in college* (2nd ed.). San Francisco, CA: Jossey-Bass.

Wieman, C. (2007). Why not try a scientific approach to science education? *Change.*http://www.changemag.org/Archives/Back%20Issues/September-October%202007/index.html

Wiggins, G. P., & McTighe, J. (1998). *Understanding by design.* Alexandria, VA: Association for Supervision and Curriculum Development.

Wood, W. B. (2009). Innovations in teaching undergraduate biology and why we need them. *Annual Review of Cell and Developmental Biology, 25,* 93–112. doi:10.1146/annurev.cellbio.24.110707.175306.

Chapter 4
Consultation for Individuals and Groups of Faculty

Many faculty members recognize the superiority of student-centered approaches over traditional teacher-centered approaches for teaching and learning, but this awareness has not translated into widespread adoption of student-centered teaching approaches (Dancy & Henderson, 2008; Henderson, Beach, Finkelstein, & Larson, 2008; Marbach-Ad, Shaefer-Ziemer, Orgler, & Thompson, 2014). Many factors impede implementation of these approaches, such as expectations of content coverage, departmental norms, and student resistance, as well as instructor time and resource constraints (Dancy & Henderson, 2008; Henderson & Dancy, 2011). Personalized consultation can help faculty members overcome these barriers to implement changes in their courses (Hativa, 1995). In the previous chapters, we described large-scale professional development activities, such as seminars and workshops, that familiarize faculty members with recommended practices. However, by its nature, such professional development programming is designed to appeal to a broad audience. Consultation complements these more generalized activities by helping faculty members overcome the idiosyncratic challenges inherent to applying recommended practices in the context of their classes. In this chapter, we describe our approach to consulting for individuals and groups of faculty, including Faculty Learning Communities (FLCs).

The Need for Ongoing Professional Development

In Chap. 3, we focused on new faculty member beliefs and practices, the challenges they face as instructors, and programming designed to meet their varied professional development needs. Based on existing literature (Austin & Sorcinelli, 2013; Sorcinelli, Austin, Addy, & Beach, 2006) and our own research (Marbach-Ad, Schaefer Ziemer, Thompson, & Orgler, 2013), we discovered that multifaceted professional development in teaching helps new faculty members overcome many of the challenges they face. However, teaching is a complex task that requires ongoing

© Springer International Publishing Switzerland 2015
G. Marbach-Ad et al., *A Discipline-Based Teaching and Learning Center*,
DOI 10.1007/978-3-319-01652-8_4

innovation. Teaching proficiency at one point in time does not constitute mastery in teaching at all times and in all settings. All faculty members, including both new and experienced faculty members, benefit from ongoing professional development for a variety of reasons:

- **Changes over time:** Faculty members at different stages of their careers may approach teaching differently. Over the course of their careers, faculty members will undoubtedly experience shifts in the distribution of time dedicated to teaching, research, mentoring, and service (Austin, 2011). Some of these shifts result from the institutional priorities that guide promotion and tenure decisions (Austin, Sorcinelli, & McDaniels, 2007; Fairweather, 2008; Lazerson, Wagener, & Shumanis, 2000).
- **Changes in teaching responsibilities:** In addition to changes in the balance between their various academic responsibilities, faculty members are likely to teach different courses over the years, which may differ in level, enrollment, student composition, and structure, as well as content. This requires faculty to adapt their teaching approaches based on the requirements of the class that they are teaching at any given point in time.
- **Changes in our knowledge of best practices:** The craft of teaching is dynamic, mainly due to continuous reforms and growth in the knowledge base on effective teaching practices.

Encouragingly, most faculty members are aware of and value student-centered, evidence-based instructional approaches (Dancy & Henderson, 2008; Henderson, Dancy, & Niewiadomska-Bugaj, 2012). However, the implementation of these practices is a complex and context-dependent task that often requires extensive human and material resources (Finelli, Daly, & Richardson, 2014; Henderson & Dancy, 2011) as well as institutional encouragement (Dancy & Henderson, 2008; Finkelstein & Pollock, 2005). Continuous professional development opportunities can help faculty close the gap between their beliefs and practices (Gibbs & Coffey, 2004; Marbach-Ad et al., 2014), which will ultimately enhance student learning (McShannon et al., 2006).

Bridging the Gap Through Professional Development

There are numerous barriers and enabling factors that impact faculty motivation and capacity for enacting instructional change. A recent University of Michigan study identified seven categories of factors that influence faculty members' adoption of evidence-based teaching practices (Finelli et al., 2014):

- **Infrastructure and culture:** Existing infrastructure and the culture in which faculty members work present barriers to implementing change in their teaching, and promote a focus on research.
- **Knowledge and skills related to effective teaching practices:** Faculty members highlighted the importance of credible and relevant research supporting any instructional practices that they might be encouraged to implement. They also

emphasized the importance of personalized support or "handholding" to support them in their implementation of techniques that were new and unfamiliar.

- **Student experience:** Faculty members were reluctant to adopt new techniques because of the risk of negative student responses or outcomes. On the other hand, student experience can be an enabler when faculty members have compelling evidence that a particular teaching approach will improve student learning outcomes and student-faculty interactions.
- **Time:** Learning about effective teaching practices and successfully implementing them can require an extensive time commitment, which presents a barrier to change.
- **Classroom and curriculum:** Faculty members varied in their levels of control over the curriculum of their courses, and raised concerns about being able to cover the required content while dedicating class time to active learning activities. They also noted barriers related to class size and the physical structure of the classroom.
- **Personal disposition:** Faculty members varied in their willingness to invest time and energy in improving their teaching.
- **Networking and community:** Colleagues were perceived as both barriers to and enablers for adopting evidence-based teaching approaches. Peer feedback on potential teaching innovations can strongly influence a faculty member's willingness to adopt the innovations.

The existence of these significant impediments suggests that bringing about widespread change requires support and professional development at multiple levels. An ongoing and comprehensive professional development program can empower faculty members to overcome the many barriers they face in changing their instructional strategies. This professional development program should include consultation. Teaching consultation is defined as support that

> ...involves dialogue between a teacher and teaching consultant...working to address a teaching concern and to improve teaching by interpreting feedback and generating improvement strategies. (Penny & Coe, 2004, p. 220)

Tagg (2010) suggested that consultation can help faculty members become self-regulating learners about their own teaching, while also breaking down the isolation of faculty members and building collaboration around a shared goal of improving their work as teachers. When offered to individuals (Hativa, 1995) and groups of faculty or departments (Kressel, Bailey, & Forman, 1999), consulting can foster widespread changes in instruction.

Our Model for Consulting

Many university teaching and learning centers offer consulting services (Sorcinelli et al., 2006). These services tend to be housed in centers that serve the whole university, sometimes with designated representatives for specific colleges or departments. Our Teaching and Learning Center (TLC) takes this specialization to another level by embedding the consultation services within the college. This decentralization brings us closer to the culture of the departments. The TLC director is a discipline-based education researcher (DBER) who specializes in the departmental disciplines. We have developed close collaborations with DBERs within each department. This network of relationships has enabled the TLC to become integrated into the culture and mission of the departments. Research suggests that DBERs can serve as effective change agents, since they understand the local environment and the context in which faculty work. DBERs have the professional expertise to cultivate change by

- Approaching teaching with the rigor employed in scientific research;
- Translating education research and theory into forms that are relevant and understandable for STEM faculty; and
- Infecting others with their enthusiasm for improving STEM education (Bouwma-Gearhart, 2012).

DBERs not only have knowledge of the body of research on effective teaching in their discipline, but also tend to be deeply involved in research on teaching within their own department. Sharing this research with departmental colleagues promotes change because evidence derived from their own department provides stronger motivation than evidence from outside institutions, which may have substantially different contexts (Graham, 2012; Hora, 2012; McKenna, Froyd, King, Litzinger, & Seymour, 2011; Singer, 2008; Wieman, Perkins, & Gilbert, 2010).

In addition to providing direct consultation services and collaborating with DBERs, we rely on the assistance of faculty with substantive scientific research backgrounds who, more recently, have become proponents of and participants in STEM education reform efforts. These faculty members can serve as "brokers" (Bouwma-Gearhart, 2012) within the department because of their interest and expertise in both science and science education. They link other faculty members to science education efforts, and serve as key recruiters and community leaders in collaborative change efforts. They build enthusiasm for reform initiatives and serve as role models who demonstrate the feasibility of creating synergies between science and science education research. As role models, faculty members who have successfully implemented evidence-based teaching approaches can demonstrate these practices to other faculty members and offer context-specific advice on how other can implement similar approaches (Finelli et al., 2014). In addition to working with these "brokers" from within our departments, we also collaborate with individuals and offices from across the campus that share our interest in teaching and learning.

While our organizational structure fosters strong connections within the department, we are independent of the departmental hierarchy. This insulates us from being involved in decisions about promotion and tenure, which makes the TLC an approachable and nonthreatening resource for faculty. This non-threatening environment serves as an enabler for faculty members who may otherwise be reluctant to share their lack of knowledge about instructional techniques and/or concerns about their own teaching (Finelli et al., 2014). Despite this formal separation, we serve as faculty advocates with respect to their teaching. Upon request, we provide letters of support and information about faculty efforts to improve their teaching in promotion and tenure reviews.

Our consultations follow a set of guiding principles and practices. First and foremost, we respect the knowledge and expertise of the faculty members we serve. They bring unique and extensive knowledge in their area of scientific research and from their teaching experience. However, they often lack pedagogical knowledge and pedagogical content knowledge, which is where our expertise lies. In consultation sessions, we come together as collaborators to share our varied expertise in pursuit of our joint mission to improve undergraduate STEM education.

Our consultations are also guided by a deliberate sensitivity to the professional realities of the faculty members we serve and the barriers they face. We tailor our consultation to fit their time availability, specific focus or interest, individual skills and characteristics, rank or position, and classroom context. We do not impose specific changes on faculty members, but rather respond to their needs and preferences. This may involve recommending specific approaches or devising an implementation and evaluation plan for an approach that they have already selected.

Our consultations generally fit into two broad—and sometimes overlapping—categories: (1) consultations for individuals; and (2) consultation for groups of faculty, including faculty learning communities (FLCs). While our consultations touch on all pedagogical content knowledge (PCK) components, they most often focus on supporting curricular and pedagogical change and developing assessment tools to evaluate student learning. Consultation can also support research on innovative teaching practices and the dissemination of this research through publications and conference presentations. Additionally, we support faculty members in writing grant proposals on topics related to teaching and learning. While our consultation service is primarily targeted at faculty members, we also invite postdoctoral fellows and graduate students to use our consulting services.

Consultation for Individuals

At the start of each semester, we contact all departmental leadership, faculty members, and graduate students via email to invite them to take advantage of our consultation services. We provide an overview of the types of consultations we offer and encourage them to contact the TLC for individualized or small group consultations, workshops, and/or teaching discussion groups. While we suggest certain topics that are likely to be of interest to our faculty, we are open and responsive to all requests for individualized support.

Sample Invitation Email for Consultation

To all faculty,

As we approach the spring semester, we would like to remind you that the TLC can offer assistance to new and experienced faculty members in course

(continued)

design and course evaluation. We can recommend innovative approaches and strategies to better engage students in their learning, bring science research into teaching, and document student understanding and satisfaction. Now is a good time to start thinking about integrating innovative technologies into your courses (e.g., clickers, wikis, social media, podcasting, and videos) as well as documenting your innovations (e.g., through pre- and post-course surveys). Many individual and group initiatives that were implemented last year have been accepted for publication in peer-reviewed journals and presented at professional meetings.

We offer individualized consultations, workshops, and teaching discussion groups. We are happy to provide advice on approaching the "broader impacts" sections of grant proposals or in preparing proposals for science education initiatives. The TLC can be of particular assistance to new or junior faculty members, who may have questions about teaching resources and approaches.

For individualized assistance, contact [TLC director]. General information on the TLC and its resources for faculty members can be found at cmns-tlc. umd.edu/tlc.

Consulting relationships usually begin as a response to faculty member requests, often in response to the emails. Requests may stem from general interests, such as a faculty member's desire to improve instruction, or specific needs, such as questions about how to implement a selected teaching approach. In other cases, the consulting relationship is initiated by a recommendation from departmental leadership. Regardless of how the relationship starts, we employ the guiding principals noted above and respond to the consultees' needs and goals.

In many cases, faculty members seek TLC consultation in response to negative student feedback on course evaluations. Research suggests that efforts to enhance instruction based on student feedback from evaluative surveys are more effective when the course instructor collaborates with a consultant (Diamond, 2004; Finelli, Pinder-Grover, & Wright, 2011; Hunt, 2003; Penny & Coe, 2004). The consultant assists the instructor by clarifying teaching goals, encouraging reflection on teaching goals and methods, observing classroom instruction, and offering feedback based on this observation to supplement feedback provided by students (Knapper & Piccinin, 1999).

Individual consultation to improve instruction generally follows a multi-step process (Finelli et al., 2011):

1. **Initial Meeting:** The consultant and the instructor meet to discuss the instructor's concerns, which most often relate to instruction, curriculum, and/or assessment. They also discuss the goals and logistics of the consultation process.
2. **Class Observation:** The consultant observes one or more class sessions and collects data through detailed field notes, an observation rubric, or video/audio recording.

3. **Student Feedback:** The consultant reviews student feedback from prior course evaluations, mid-term course evaluations from the current course, informal feedback collected by the instructor, and/or feedback collected by the consultant through interviews, focus groups, or surveys.
4. **Summary of Findings and Recommendations:** The consultant prepares a summary of data collected through classroom observation and student feedback, then develops recommendations based on these findings. The consultant may share these findings and recommendations through a report, scored rubric, and/or face-to-face meeting. Summaries and recommendations remain confidential except in the case of an explicit request by the consultee. Upon such a request, the consultant may provide a summary or report to departmental leadership and/or a tenure review committee.
5. **Follow-Up:** Upon instructor request, the consultant will provide ongoing feedback and support for the implementation of recommendations. Follow-up varies widely based on the needs and goals of the consultee.

While these steps illustrate the typical individual consultation process, our consultation services vary widely in form and focus depending on the consultee's needs. The Implementation Guide gives an overview of the variety of individual consulting services available from the TLC. We have categorized our consultations based on the five PCK components, and provide one or two examples for each component. For more information about each PCK component, see Chap. 1.

Consultation for Groups of Faculty

In addition to consultation for individuals, the TLC offers consultation for groups of faculty. These group consultations generally fit into one of two categories: consultations for informal groups of faculty and consultation for formal FLCs.

Informal Groups of Faculty

Informal groups of faculty generally consist of a few instructors who teach parallel or closely related courses. These consulting relationships typically resemble those described in the preceding section on individual consultation, but working as a group allows for change that simultaneously impacts courses taught by multiple instructors.

One such consultation occurred when a group of three faculty members came to the TLC for assistance in redesigning a large, team-taught introductory biology class. Each faculty member taught one third of the material in two different lecture sections of the course. The group wanted to try alternative instructional approaches in the two sections, and sought assistance in designing a comparative study to assess the relative effectiveness of the newly implemented teaching methods. Two of the three weekly class sessions would be the same for both sections in terms of teaching approach and content. For the third class session, each section would focus on the same topic, but use a different teaching approach: traditional lecture with embedded demonstrations in one section, active learning in the other. The active learning section was designed to incorporate small-group, active learning engagement exercises with undergraduate learning assistants circulating throughout the class to assist students in completing the exercise.

TLC consultation for the group involved assistance with writing a grant proposal to fund the development of new course materials, selecting assessment tools to measure the effectiveness of the two approaches, collecting and analyzing data, and disseminating the study findings in conferences and peer-reviewed publications. The TLC director also provided feedback on ways to implement the different instructional approaches and tips for improving their effectiveness. As a result of the consultation process, the group received funding and is currently engaged in carrying out this study.

Faculty Learning Communities

The TLC also provides consultation services for FLCs, which serve an important role in our comprehensive professional development program. Our conceptualization of FLCs is derived from Wenger's (1998, p. 1) theory of communities of practice. As Wenger (http://wenger-trayner.com/wp-content/uploads/2012/01/06-Brief-introduction-to-communities-of-practice.pdf) explains, "Communities of practice are groups of people who share a concern or a passion for something they do and learn how to do it better as they interact regularly." In our university context, we define FLCs as groups of faculty who meet on a regular basis and share activities, knowledge, and practices. While faculty within FLCs may collaborate on a variety of enterprises, in this book we focus only on those communities whose joint enterprise involves improving undergraduate science education.

FLCs vary in their lifespans. Some of our FLCs are active for a year or so, while others are longstanding. For example, the Host Pathogen Interaction (HPI) community (Marbach-Ad et al., 2007), one of our most prominent FLCs, has been continuously active for more than 10 years.

Our FLCs generally include individuals of different faculty ranks and positions, with a combination of tenure-track and non-tenure-track faculty members, as well as postdocs and graduate students. Most of our FLCs include one or more DBERs whose content area or research interest is closely connected to the shared enterprise of the group. DBER participation grounds the group in science education research and practice, and provides pedagogical content knowledge. In recent years, discipline-based FLCs with DBER participation have become increasingly common because of the key role of disciplinary cultures and PCK in education reform (Marbach-Ad et al., 2007).

Types of FLCs

Our FLC are organized around different foci, including (a) gateway introductory courses, (b) the interface between related disciplines, and (c) courses related to a shared research area. We also have (d) a community targeted to non-tenure-track faculty members that meets for monthly discussions.

(a) **Communities built around gateway courses**

One type of FLC involves the instructors responsible for teaching a specific gateway course, as well as instructors teaching courses that are taken before or after the gateway course. These FLCs generally aim to address issues related to pedagogy (i.e., instituting best practices for effective, student-centered teaching) and/or issues related to content (i.e., improving linkages with courses generally taken before and after the gateway course). One exemplar of this type of FLC is the Cell Biology and Physiology Working Group, which focuses on improving student learning in a gateway cell biology course (see text box for more details).

An FLC Built Around a Gateway Course: The Cell Biology and Physiology Working Group

Approximately 15 faculty members from the biology and cell biology departments, as well as faculty from a satellite campus, created a community to increase continuity between the Cell Biology and Physiology course and related upper-level courses. This course, which is typically taken in the second or third year of the biological sciences curriculum, is a prerequisite for all upper-division cell biology and physiology courses and serves a large number of students in biological sciences and bioengineering majors. One of the goals of this FLC was to take a learning progression approach to developing course content. This approach involved assessing the knowledge that students bring to the Cell Biology and Physiology course, and the knowledge that they take from this course to upper-level courses. The FLC developed a concept inventory that could be used to measure student learning after completion of each course in the sequence. Additionally, the group revised the laboratory portion of the course to make it more research-like, with open-ended investigations that utilize modern research methodologies and instrumentation.

(b) **Communities for interdisciplinary collaboration**

Other FLCs focus on the interface between biology or chemistry and related disciplines, as the following examples illustrate. The mathematics/biology FLC has resulted in the development of a new calculus sequence specifically designed for biology students, a new upper-level mathematical biology course, and 42 MathBench online interactive modules that teach introductory biology students how to apply quantitative approaches to biological problems (Nelson, Marbach-Ad, Thompson, Shields, & Fagan, 2009; Thompson et al., 2013). A physics/biology FLC has enriched introductory biology courses by placing greater emphasis on interdisciplinary connections between biology and physics, such as how organisms respond to constraints imposed by the physical environment. This collaborative effort has also spurred revisions to the physics sequence taken by biology majors (Redish et al., 2014). A third FLC focused on the chemistry/biology interface and developed a linked set of laboratory exercises in which silver nanoparticles developed by students enrolled in an organic chemistry course were tested for antibacterial properties by students enrolled in general microbiology.

(c) **Communities built around research areas**

A third type of FLC involves groups of faculty that share a common research focus and teach courses related to their area of research. An example of this type is the Host Pathogen Interaction (HPI) teaching group, which is comprised of approximately 20 faculty members from the Department of Cell Biology and Molecular Genetics. They have worked over the past 4 years to improve eight linked microbiology and immunology courses (see text box for more details).

A Community Organized Around a Shared Research Area: The Host-Pathogen Interaction FLC

This community brings together 20 faculty members with a shared research interest in HPI who are collectively responsible for nine undergraduate courses that cover HPI topics. The HPI FLC has met monthly since 2004 to discuss topics related to teaching and learning. Some faculty members have been continuously involved in the community since 2004, while others have joined more recently. The FLC includes tenure-track and non-tenure track faculty members of varying experience levels. Postdoctoral fellows and graduate students interested in the interface between HPI research and science education also have participated in FLC activities.

The HPI community has carried out a number of successful initiatives over the last 10 years. For example, they have developed a concept inventory that has been used to coordinate the coverage of HPI topics across nine courses. Their collaboration has also resulted in a number of science education grants, conference presentations, and peer-reviewed journal publications. For a more detailed description of this community and its work, see Marbach-Ad et al. (2009) and www.hpiresearchteachingteam.umd.edu.

...[HPI community discussions] about phenotype/genotype opened my eyes to students' understanding, especially in [their] senior year. Only 30 % grasp the whole picture. Students compartmentalize their knowledge between courses: 'What happens in Vegas stays in Vegas.'

(continued)

> *...I have learned a great deal about pedagogy strategies from the HPI teaching team. I am also gaining insight into expectations about student knowledge/conceptual understanding across various HPI and Cell Biology courses.*
>
> –HPI FLC members

(d) Communities targeted to specific groups of participants

One of our communities is not organized around a specific mission but instead brings together non-tenure-track lecturers to discuss teaching and learning issues that are of particular relevance to them. Within our college, lecturers are distributed through several departments, and each department employs only a relatively small number of them. In this context, lecturers may sometimes feel that departmental discussion is skewed in favor of tenure-track faculty concerns. Lecturers also have responsibilities that differ from those of their tenure-track colleagues. For example, they are likely to have heavier teaching loads, more involvement in teaching large-enrollment lecture courses, responsibility for coordinating multiple laboratory and/or discussion sections that accompany lecture courses, and responsibility for providing professional development in teaching to graduate and undergraduate teaching assistants. These different professional responsibilities can lead to different professional development concerns, needs, and preferences, which this type of FLC aims to address. The lecturer FLC meets a few times a semester, often over lunch. Although this FLC is designed to meet the specific needs of our lecturers, tenure-track faculty are welcome to participate on a regular basis or on occasions when the topic of discussion is of particular interest.

Benefits of FLCs

FLCs foster productive collaboration among diverse participants who might otherwise work in isolation (Cox, 2004). They provide an ongoing opportunity for less experienced faculty members to be mentored by more senior faculty members (Ash, Brown, Kluger-Bell, & Hunter, 2009; Sorcinelli, 1988). Through FLC interactions, less experienced faculty members gain confidence and become acculturated into their teaching roles.

FLC participants gain familiarity with and derive support for the adoption of effective teaching approaches (Dawkins, 2006; Lakshamanan, Heath, Perlmutter, & Elder, 2011; Layne & Froyd, 2006; Silverthorn, Thorn, & Svinicki, 2006; Sirum, Madigan, & Klionsky, 2009; Vescio, Ross, & Adams, 2008). Cox (2004) found that faculty participating in FLCs reported more favorable classroom environments, more student engagement, and higher levels of student critical thinking than faculty who were not participating in FLCs. Our experience indicates that faculty members who participate in communities feel more collegial and more recognized for their teaching than faculty members who do not participate in FLCs (Marbach-Ad et al., 2007).

The teamwork that develops within a community creates a supportive environment that promotes faculty exploration and growth in all components of PCK. Ongoing teamwork not only benefits individual participants but also makes it more feasible for groups of faculty to engage in and obtain grant funding for large-scale, complex initiatives.

> ... *This community gets me thinking about ways to make my teaching more interesting and more effective. I get ideas that I don't get any other place.*
>
> *I gain ideas that I can implement in my classes and share with colleagues. They also provide synergistic interactions and brainstorming opportunities that often result in grant proposals to further our efforts.*
>
> ... *I enjoy EVERY meeting because I always learn something new.*
>
> ... *As a researcher who teaches, I learn about the field of science education and current approaches to improve learning and literacy... Also, [the community] allows for publications in education that otherwise would never be pursued due to time constraints in running a research lab.*
>
> ... *It gives me a sense that the department cares about teaching and improving teaching.*
>
> ... *[I benefit from] hearing about new developments in teaching and learning, discussing new research that may influence my teaching, giving feedback to other faculty who are trying new things or trying to assess the impact of teaching innovations, getting motivated!*
>
> –Faculty members on the benefits of participating in TLC-supported FLCs

The Role of the TLC in FLCs

The TLC's consulting services for FLCs are ongoing and holistic. TLC support may vary throughout the life of the FLC, based on the needs of the community and the specific activities it undertakes. Our support generally touches on all of the PCK components as FLCs engage in different activities over time. In the Implementation Guide, we describe our consultation and support for groups of faculty members related to the five PCK components.

Conclusion

Our consulting services complement the full range of TLC activities. While consultations are a time- and labor-intensive form of professional development, in many cases they are the most impactful. Whether the consultation serves an individual or a group, consultations allow us to tailor our support to the specific needs, teaching conditions, and interests of the faculty members we serve. By providing consultation services that are specifically targeted to the chemistry and biology departments, we have built strong relationships and credibility within this community of faculty members and departmental leadership. Our discipline-specific focus allows us to rely on DBERs who have strong expertise in science education and speak the language of the faculty members we serve.

Implementation Guide

The Implementation Guide for this chapter provides examples of consultations for individuals and for groups that address the different PCK components. While requests for consulting in teaching will vary across settings, these examples illustrate the breadth of TLC consulting services and the role of highly personalized consulting in creating a culture that fosters innovation in teaching.

Consultation for Individuals

Below, we describe our individual consulting service in terms of how it addresses the different components of PCK. For each component, we provide examples that illustrate how consulting can promote instructional change. Pseudonyms are used to protect the confidentiality of the faculty members.

PCK 1: Knowledge of Student Understanding

Our individual consulting related to knowledge of student understanding of science generally consists of helping instructors determine methods to better understand student backgrounds. Student background characteristics include demographics, majors, prior coursework, prior knowledge, interests, and preferred learning styles. Methods to collect information on these characteristics include surveys, student information system data, and informal conversations with students. We also assist instructors in adjusting their instruction to reflect student diversity using methods that are appropriate to the course and its content.

Jenna, a biology instructor, learned about Felder's Index of Learning Styles at the welcome workshop for new instructors (see Chap. 3 for a more detailed description of this workshop). She came to the TLC seeking assistance with using this index to develop an awareness of the student learning styles in an interdisciplinary course that she was slated to teach. TLC staff provided theoretical background underpinning Felder's learning style theory, suggestions for how to interpret data from the index, and advice for how to use these data to cater instructional techniques for the diverse population of students in the class. Jenna explained how she ultimately used this information:

> I do a unit on the learning styles in [my class] and we talk about it in the beginning . . . I gather the data anonymously for the class in terms of who's verbal, who's visual, what fraction are sensing and intuiting. I think that it's always helpful to be thinking about what kind of tools will help this class learn the material.

Shortly thereafter, Jenna applied for and received a grant from the National Science Foundation (NSF). In her grant application, which required a teaching innovation component, she highlighted her use of the Felder Learning Style Index to shape her instructional approaches. TLC staff assisted her in the preparation of the teaching component of the grant application and provided a letter of support.

PCK 2: Knowledge of Science Curriculum

Individuals who come to the TLC seeking consultation and support centered on this PCK component generally aim to address curriculum issues within a specific course that they teach. They may seek advice on sequencing topics within that course, defining learning goals for key course topics, or building the course around broad principles or concepts.

Sometimes, making curricular changes to a single course presents challenges due to the interconnectedness of that course with other courses (e.g., large-enrollment courses with multiple sections or sequences of courses). Making changes in such situations requires collaboration of multiple faculty members. In some cases, we refer the faculty member to an existing group or community that addresses related curricular issues. In other cases, we have suggested that a new group be formed to address curricular changes.

Omar taught a biology class for non-majors. Because this course did not fit into a defined sequence of courses, he had a lot of flexibility in terms of the topics that the course covered. Over a few semesters, Omar noted that his students did not seem interested in topics he taught, and he came to the TLC seeking assistance in selecting topics that would be more interesting to his students. We suggested guidelines for selecting topics, such as using topics that relate to everyday life and the students' surroundings. We also discussed how the selection of topics impacts the range of possible instructional strategies. For example, by incorporating topics related to plants native to Maryland, Omar was able to take his students out into the field to collect samples as part of a class activity.

Dana, a lab coordinator in an introductory biology course, sought to revamp her lab and came to the TLC for consultation. She was limited in the scope of change that she could implement because her lab was connected to a lecture course taught by multiple faculty members. Furthermore, this lecture and lab combination served as a gateway to many upper-level courses. With the assistance of the TLC, Dana initiated a new community that included the lecture course instructors, as well as the instructors of related upper-level courses. The community worked together to determine which topics Dana's lab should cover to support the lecture course and to build a knowledge progression across the related courses. Dana was then able to revamp the course while still covering the material that students needed for success in the lecture portion of the course and subsequent courses.

PCK 3: Knowledge of Instructional Strategies

Our consultations in this PCK category have ranged from specific requests for assistance (e.g., how to use clickers to engage students in a large enrollment course) to broad requests for help with ambitious projects (e.g., how to revise a course to implement blended learning). These consultations have sometimes involved a combination of direct assistance and referrals to outside resources. For example, in assisting faculty members with incorporating clickers, we have worked with faculty members to develop cognitively challenging clicker questions, but referred them to a specialist in instructional technology for technical assistance with implementation. We often refer faculty members to their peers within the College who have successfully implemented the same instructional strategies.

Maria, a chemistry instructor, sought help fostering student participation in the large-enrollment introductory course she taught. While she frequently asked students to participate in class discussions, she was unsatisfied with the level and quality of their participation and did not know how to improve this.

Maria met with the TLC staff to explore her options. In the meeting, we discussed Maria's class setting, a large auditorium, and the approaches she was using to encourage student participation. We suggested techniques for more effectively eliciting student responses to her questions, such as allowing more time for students to formulate their answers and providing positive feedback to build a supportive atmosphere for students who might be unsure of the correctness of their responses.

We also suggested a variety of structured student engagement techniques that are feasible in large classes and can reduce the intimidation that students may feel about responding in those settings (e.g., think-pair-share, one-minute essays, and clicker questions). We discussed the pros and cons of assigning grades for student participation and having students volunteer to respond versus calling on specific students. We also talked about the importance of allowing students multiple avenues for participation, such as having a mix of individual, small group, and whole class work. We provided Maria with scholarly papers about the various techniques, and referred her to other faculty members who were using these techniques in their own classes.

Maria decided to incorporate the think-pair-share technique and to work on fostering a supportive environment for student participation in class discussions. A few weeks after our initial meeting, we met with Maria to follow up on the degree and quality of student participation in her class. She reported that student participation had increased noticeably, and that students seemed very engaged in the think-pair-share activities. She also commented that having the students work in pairs had the additional, unanticipated benefit of allowing her to circulate within the class to gain a better real-time understanding of students' comprehension of course material.

PCK 4: Knowledge of Assessment

One of the most common areas of consultation concerns issues related to assessment. Many faculty members find this to be a particularly challenging aspect of teaching, in part because they received little or no training in this area during their graduate education.

Consultations in this component cover how to assess student learning for the purpose of assigning grades and to provide feedback to both student and instructor. Consultations may also address how to evaluate the impact of a new instructional technique or strategy. Faculty members generally come to us seeking assistance in determining which assessment tools they should use for different purposes. Once they decide upon a particular type of assessment, we offer them assistance in writing assessment items, implementing the assessment, and analyzing the results.

Donna, a biology instructor, came to the TLC for help in addressing negative student feedback about her course's final exam. In course evaluations and face-to-face interactions, her students had indicated that they loved her course but were very frustrated by what they considered an "unfair" final exam. Donna noted that she lacked experience in writing exam items. During her consultation, we reviewed the exam questions and the pattern of answers selected by students on the multiple-choice exam. Students did not perform poorly on the exam as a whole, but an analysis of responses indicated that nearly all students chose incorrect answers for three particular items. When we looked more closely at these items, we found them to be ambiguous and confusing, which was probably the root cause of students' poor performance and frustration.

After discussing this issue, we suggested various ways to ensure that all items on future exams are clear and reliably assess student knowledge of the target content. We recommended some specific quality control checks, such as having graduate teaching assistants review all test items prior to giving them to students and using a combination of item types (i.e., open-ended questions as well as multiple choice questions). The TLC director also gave Donna a resource on how to develop assessment tools and characteristics of strong items (Walvoord & Anderson, 2010).

Trevor, a chemistry instructor, recently developed a series of 12 videos that guided students through the course textbook's end-of-chapter problems. This additional resource was recommended to students but not required. Trevor came to the TLC seeking a method to assess the impact of the videos on student learning. Using the university's online learning management system, he was able to track when and how often each student watched each video, which provided a wealth of raw data.

We helped Trevor to design and conduct an evaluation of the impact of the videos. The evaluation included two components. First, we conducted an analysis of the effect of accessing the videos on student exam grades, with cumulative GPA as a covariate. Then, we developed a survey with open-ended questions designed to gauge students' perceptions of the effectiveness of the videos and the ways in which the videos contributed to their learning. These qualitative data were particularly helpful in identifying the situations in which students found the videos most helpful.

Trevor presented the results from this evaluation at several conferences. The TLC supported him in developing conference proposals and presentation materials, as well as in providing travel grants to defray the costs of his travel to the conferences.

> *The online video lectures are extremely helpful... When approaching a chemistry problem, one of the hardest things to do is to find out where to start and what are the subsequent steps after you start... I have always tried problems on my own with only the help from the book... I would usually get frustrated when I would get the answer wrong. This frustration would result in a loss of confidence... [Now] I can supplement the book and class notes with these online video lectures, where there is a person to not only help me get the right answer but [who] would also explain the fundamental logic behind each step. So I could find out where I went wrong and would correct that mistake. This exact situation played out throughout the entire semester and made me a better student with stronger intellectual understanding of chemistry.*
>
> —Student feedback on video evaluation survey

PCK 5: Orientation to Science Teaching

Faculty who seek consultation related to this PCK category usually come to the TLC for help in developing teaching philosophies and to learn about science education theories. Many seek this assistance in the context of writing sections of grant proposals, such as the science education component of NSF CAREER grants or broader impact statements for other NSF grants.

Linda, a biology faculty member, came to the TLC for assistance in developing a teaching philosophy statement to enhance her attractiveness as a candidate for a faculty position for which she hoped to apply. The TLC director met with her and shared examples of faculty teaching philosophy statements, as well as prominent science education research and policy reports that could serve as resources in the preparation of her statement (AAAS, 2011; Handelsman, Miller, & Pfund, 2007). After the meeting, Linda sent a revised draft of her teaching philosophy statement to the TLC for review. The TLC director and another faculty member provided feedback to strengthen her teaching philosophy statement.

Consultation for Faculty Learning Communities

In the pages that follow, we describe the consulting services that we have provided for FLCs, organized by the five PCK components. The examples provided illustrate the breadth and scope of instructional change that the consultation process can facilitate. Additionally, the examples illustrate how complex activities undertaken by FLCs (e.g., developing and using a concept inventory) require integration of knowledge across multiple PCK components.

PCK 1: Knowledge of Student Understanding

When we offer consultation to FLCs for this PCK component, we generally support them in developing mechanisms to better understand students' prior knowledge and knowledge development. In many cases, this relates to how students retain knowledge across a sequence of courses.

Groups of faculty members can address students' prior knowledge and knowledge development more comprehensively than individual faculty members for three primary reasons. First, involving the instructors of all the courses in the sequence makes it possible to do coordinated assessment of student learning along the entire progression. Second, each instructor brings unique insight to diagnosing learning difficulties and devising interventions, which increases the likelihood that the interventions developed by the group will be successful. Third, having all the relevant instructors involved makes it more feasible to mount a coordinated response across all courses in the sequence.

FLCs can work collaboratively to collect information about what was taught and learned in prior courses. With this information, they can address gaps between the knowledge that students bring to a class and the knowledge that the instructor expects them to have. Our consultations for FLCs often involve developing tools to probe students' prior understanding and any alternative conceptions that they may hold. In many cases, multiple courses in a sequence use a single tool to measure the progression of student learning and the impact of early interventions on achievement in subsequent courses.

A chemistry FLC includes faculty members responsible for teaching the four courses in our general and organic chemistry sequence. A major challenge facing the instructors of the second course in this sequence was the diversity in student backgrounds. Students reached this course via different paths: some had taken the prerequisite course at our university, others had completed a course deemed to be equivalent in high school, a community college, or a different four-year university. Course instructors reported that students come to the course with varying degrees of mastery of prior content. This diversity in background knowledge caused those with weaker backgrounds to struggle in the second course.

To address this challenge, the FLC sought to develop an instrument to assess students' previous knowledge of key topics. The TLC assisted the group in refining an existing assessment tool. They worked together to determine what topics this instrument would probe and to develop a plan for implementing and using the instrument. Once the tool had been implemented, the TLC continued to support the FLC by assisting with the analysis of student responses and providing guidance on appropriate instructional responses based on this analysis. Additionally, the TLC assisted the group in securing a grant to support their initiative.

In another instance, several faculty members who taught upper-level courses in genomics noticed that students struggled to understand several concepts that were predicated on a basic understanding of gene structure and function. They discussed their concerns with TLC staff, who then convened an FLC. The newly formed FLC consisted of the upper-level instructors, as well as colleagues who taught the prerequisite introductory biology and principles of genetics courses in which the problematic concepts were first introduced. The instructors of the prerequisite courses were surprised by the difficulties that students exhibited in the upper-level courses, because they had covered the relevant background content repeatedly. The FLC decided to implement a common diagnostic assessment across the sequence of courses. This assessment focused on biology's Central Dogma (DNA codes for RNA, which codes for proteins). It required students to interpret diagrams of this process, and draw their own version of the process given a basic template.

This assessment revealed that students generally were adept at correctly answering multiple-choice questions on the topic, but their understanding of the dynamic process was uneven. A postdoctoral science education researcher conducted a detailed follow-up study of a related genetic process (DNA replication) and discovered the same deficiencies in moving from knowledge of basic terms and concepts to a deep understanding of the physical and dynamic aspects of the process.

As an intervention, members of the FLC created several different worksheets based on the original assessment. These worksheets gave students at all levels of the curriculum additional practice at conceptualizing the Central Dogma. Informal reports from the upper-level faculty indicate that students are now showing better comprehension of the formerly problematic concepts.

PCK 2: Knowledge of Science Curriculum

FLCs are particularly appropriate settings to address issues related to curriculum. Because of the interdependence of courses, groups of faculty teaching related courses should collaborate in making decisions that impact how students build knowledge and conceptual understanding across courses. The TLC supports faculty in designing their curricula to enhance students' development of conceptual understanding over multiple courses.

The nature of the FLC, and the range of topics taught by FLC members, impacts the complexity of this task. Some FLCs cover a broad range of topics or even interdisciplinary topics. One of the initial tasks in most FLCs is determining the key concepts that students should master in each of the courses addressed by the FLC.

Shortly after its formation, the HPI FLC set out to determine what HPI concepts they wanted students to retain 5 years after graduation. As a group, they decided on 13 key concepts. They used Allen's (2003) Curriculum Alignment Matrix to map when and to what degree these concepts were addressed in the nine HPI-related courses. This mapping revealed some gaps in coverage of the concepts. The group worked together to remedy these gaps by adjusting curriculum, with the goal of creating a learning progression that will enable students to master all of the key HPI concepts over the course of their undergraduate degree program (Marbach-Ad et al., 2010).

PCK 3: Knowledge of Instructional Strategies

Another common task of FLCs is implementing more effective instructional strategies. Some FLCs form just for this purpose. Such FLCs often focus on specific instructional strategies (e.g., blended learning, inquiry-based teaching). Other FLCs consider issues related to instructional strategies as part of a broader mission. A well-functioning FLC creates an atmosphere in which faculty learn more about innovative teaching methods, brainstorm with colleagues about the best ways of implementing those methods, and get feedback on the challenges of implementation.

The TLC supports FLC activities related to this PCK component by pointing FLCs to appropriate resources, including publications, websites with teaching materials, and on- and off-campus conferences on teaching and learning. The TLC staff also gives guidance on choosing and implementing evidence-based teaching approaches that are likely to be successful in the courses in question. Additionally, the TLC encourages FLCs to conduct research on the effectiveness of newly implemented teaching approaches and disseminate their findings through conference presentations and journal articles.

The HPI FLC developed different instructional methods to convey effectively the key concepts that the group identified as learning objectives across the nine HPI courses. The group focused on developing interactive modules, with many of these modules involving a case study approach. The case studies took different forms in different courses, with some using cases that are relevant to everyday life, some using cases derived from scientific research, and some combining these two approaches (Quimby, McIver, Marbach-Ad, & Smith, 2011). These case studies challenged students to engage in a scientific problem-solving process. In the case studies, students simulated many aspects of the research process, including review of relevant literature, data collection, data analysis, and written and oral presentation of research findings. By introducing the case study approach in introductory-level courses and continuing its use in related upper-level courses, students became familiar with the problem-solving process and developed progressively more sophisticated critical thinking skills.

The implementation of case studies across multiple courses was part of a broader strategy to implement innovative and evidence-based practices. The FLC also implemented other new instructional strategies, including group work, concept mapping, and just-in-time teaching, in their courses (Cathcart, Stieff, Marbach-Ad, Smith, & Frauwirth, 2010; Injaian, Smith, German Shipley, Marbach-Ad, & Fredericksen, 2011; Senkevitch, Marbach-Ad, Smith, & Song, 2011).

PCK 4: Knowledge of Assessment

Groups of faculty members working together constitute a powerful mechanism for creating and validating assessment tools, as well as analyzing and interpreting the resultant data. Group members bring to the table diverse expertise, perspectives, and experience with different courses. FLCs can collaborate to develop and validate an assessment tool. Additionally, by working in a group, tasks that would be onerous for an individual can be subdivided among multiple people.

Given the complexity associated with assessing student learning, the TLC staff bring not only their collective knowledge and experience, but also connections to assessment experts from across the university and other universities. The TLC has introduced FLCs to existing, validated assessments tools and provided guidance in the validation of assessment instruments developed by the FLC. As FLC participants introduce assessment tools or techniques in their courses, TLC staff assist in the analysis and interpretation of assessment results. These results are then used to guide revisions to instruction or curriculum.

One of the most comprehensive and intensive efforts related to student assessment comes from the HPI FLC. As was previously discussed, this FLC identified 13 key concepts to be covered in the nine HPI courses. In order to characterize the progression of student learning in these key concepts, the group developed an 18-item concept inventory. The concept inventory consisted of multiple-choice questions. The group engaged in an iterative process to design items with distractors that reflected commonly held alternative conceptions. Following each multiple choice item, students were asked to explain their answer in open-ended format. The open-ended responses informed the revision and validation of the concept inventory.

The group invested a great deal of time and energy in the analysis of the multiple choice and open-ended responses (Marbach-Ad et al., 2009, 2010). By looking at the pattern of correct multiple choice responses from pre- and post-tests across a series of courses, the faculty members were able to see where in the sequence of courses students gained or lost understanding of key concepts. The group analyzed the open-ended items to better understand how students think about key concepts and uncover their alternative conceptions. Having a large group facilitated this analysis of qualitative data from a large sample of students. Single faculty members or small groups of faculty members generally have limited time to engage deeply in such activities, but the larger group was able to divide the questions between small groups to increase the efficiency of data analysis without sacrificing validity.

Working on the concept inventory also enabled faculty members to be more effective in developing assessments for their own classes. As one faculty member noted, "Analyzing [concept inventory] questions according to Bloom taxonomy helped me while formulating my exams." Additionally, in the process of working on assessment tools, communities often find themselves addressing other PCK components. HPI FLC members indicated that one of the most valuable contributions of the concept inventory was their enhanced awareness of students' prior knowledge and how students carry knowledge across related courses (PCK 1).

PCK 5: Orientation to Science Teaching

FLCs tend to address this component of PCK indirectly. The collaborative work of the community, and the integration of science education specialists and DBERs in FLCs, promotes a scientific approach to science teaching. Through their collaborative activities, FLC members gain an increasing awareness of the large body of literature on effective, evidence-based practices in science education and become comfortable using the language of the field. In this process, FLC members become more reflective science instructors who see themselves as members of a community of practice.

References

Allen, M. J. (2003). *Assessing academic programs in higher education.* New York, NY: Jossey-Bass.

American Association for the Advancement of Science (AAAS). (2011). *Vision and change: A call to action.* Washington, DC: AAAS.

Ash, D., Brown, C., Kluger-Bell, B., & Hunter, L. (2009). Creating hybrid communities using inquiry as professional development for college science faculty. *Journal of College Science Teaching, 38*(6), 68–76.

Austin, A. E. (2011). *Promoting evidence-based change in undergraduate science education.* A paper commissioned by the National Academies National Research Council Board on Science Education. http://dev.tidemarkinstitute.org/sites/default/files/documents/Use%20of%20Evidence%20in%20Changinge%20Undergraduate%20Science%20Education%20%28Austin%29.pdf

Austin, A. E., & Sorcinelli, M. D. (2013). The future of faculty development: Where are we going? *New Directions for Teaching and Learning, 133,* 85–97.

Austin, A. E., Sorcinelli, M. D., & McDaniels, M. (2007). Understanding new faculty: Background, aspirations, challenges, and growth. In R. Perry & J. Smart (Eds.), *The scholarship of teaching and learning in higher education: An evidence-based perspective* (pp. 39–89). Dordrecht, Netherlands: Springer.

Bouwma-Gearhart, J. (2012). Engaging STEM faculty while attending to professional realities: An exploration of successful postsecondary STEM education reform at five SMTI institutions. *APLU/SMTI paper 5*. Washington, DC.

Cathcart, L. A., Stieff, M., Marbach-Ad, G., Smith, A. C., & Frauwirth, K. A. (2010). *Using knowledge space theory to analyze concept maps*. Paper presented at the 9th International Conference of the Learning Sciences.

Cox, M. D. (2004). Introduction to faculty learning communities. *New Directions for Teaching and Learning, 97*, 5–23.

Dancy, M., & Henderson, C. (2008). *Barriers and promises in STEM reform*. Paper presented at the commissioned paper for National Academies of Science Workshop on linking evidence and promising practices in STEM undergraduate education, Washington, DC.

Dawkins, P. W. (2006). Faculty development opportunities and learning communities. In N. Simpson & J. Layne (Eds.), *Student learning communities, faculty learning communities, & faculty development* (pp. 63–80). Stillwater, OK: New Forum.

Diamond, R. M. (2004). The usefulness of structured mid-term feedback as a catalyst for change in higher education classes. *Active Learning in Higher Education, 5*(3), 217–231.

Fairweather, J. (2008). *Linking evidence and promising practices in science, technology, engineering, and mathematics (STEM) undergraduate education: A status report for the National Academies National Research Council Board on Science Education*. Commissioned paper for the National Academies Workshop: Evidence on Promising Practices in Undergraduate Science, Technology, Engineering, and Mathematics (STEM) Education.

Finelli, C. J., Daly, S. R., & Richardson, K. M. (2014). Bridging the research-to-practice gap: Designing an institutional change plan using local evidence. *Journal of Engineering Education – Special Issue on the Complexities of Transforming Engineering Higher Education, 103*(2), 331–361.

Finelli, C. J., Pinder-Grover, T., & Wright, M. C. (2011). Consultations on teaching. Using student feedback for instructional improvement. In C. Cook & M. Kaplan (Eds.), *Advancing the culture of teaching at a research university: How a teaching center can make a difference* (pp. 65–79). Herndon, VA: Stylus.

Finkelstein, N. D., & Pollock, S. J. (2005). Replicating and understanding successful innovations: Implementing tutorials in introductory physics. *Physical Review Special Topics-Physics Education Research, 1*(1), 010101.

Gibbs, G., & Coffey, M. (2004). The impact of training of university teachers on their teaching skills, their approach to teaching and the approach to learning of their students. *Active Learning in Higher Education, 5*(1), 87–100.

Graham, R. (2012). *Achieving excellence in engineering education: The ingredients of successful change*. London, UK: The Royal Academy of Engineering.

Handelsman, J., Miller, S., & Pfund, C. (2007). *Scientific teaching*: W.H. Freeman & Company in collaboration with Roberts & Company Publishers.

Hativa, N. (1995). The department-wide approach to improving faculty instruction in higher-education: Qualitative evaluation. *Research in Higher Education, 36*(4), 377–413.

Henderson, C., Beach, A., Finkelstein, N., & Larson, R. S. (2008). *Preliminary categorization of literature on promoting change in undergraduate STEM*. Paper presented at the Facilitating Change in Undergraduate STEM symposium, Augusta, MI.

Henderson, C., & Dancy, M. (2011, February 7–8). *Increasing the impact and diffusion of STEM education innovations*. Paper presented at the National Academy of Engineering Forum – The impact and diffusion of transformative engineering education innovations, New Orleans, LA.

Henderson, C., Dancy, M., & Niewiadomska-Bugaj, M. (2012). The use of research-based instructional strategies in introductory physics: Where do faculty leave the innovation-decision process. *Physical Review Special Topics - Physics Education Research, 8*(2), 020104.

Hora, M. T. (2012). Organizational factors and instructional decision-making: A cognitive perspective. *Review of Higher Education, 35*(2), 207–235.

Hunt, N. (2003). Does mid-semester feedback make a difference? *The Journal of Scholarship of Teaching and Learning, 3*(2), 13–20.

Injaian, L., Smith, A. C., German Shipley, J., Marbach-Ad, G., & Fredericksen, B. (2011). Antiviral drug research proposal activity. *Journal of Microbiology & Biology Education, 12*, 18–28.

Knapper, C., & Piccinin, S. (1999). Consulting about teaching: An overview. *New Directions for Teaching and Learning, 79*, 3–7.

Kressel, K., Bailey, J. R., & Forman, S. G. (1999). Psychological consultation in higher education: Lessons from a university faculty development center. *Journal of Educational and Psychological Consultation, 10*(1), 51–82.

Lakshamanan, A., Heath, B. P., Perlmutter, A., & Elder, M. (2011). The impact of science content and professional learning communities on science teaching efficacy and standards-based instruction. *Journal of Research in Science Teaching, 48*, 534–551.

Layne, J., & Froyd, J. (2006). Faculty learning communities: Engaging faculty on the topic of learning. In N. Simpson & J. Layne (Eds.), *Student learning communities, Faculty learning communities, & faculty development* (pp. 81–102). Stillwater, OK: New Forum.

Lazerson, M., Wagener, U., & Shumanis, N. (2000). Teaching and learning in higher education, 1980–2000. *Change, 32*, 12–19.

Marbach-Ad, G., Briken, V., El-Sayed, N., Frauwirth, K., Fredericksen, B., Hutcheson, S., ... Smith, A. C. (2009). Assessing student understanding of host pathogen interactions using a concept inventory. *Journal of Microbiology and Biology Education, 10*, 43–50.

Marbach-Ad, G., Briken, V., Frauwirth, K., Gao, L. Y., Hutcheson, S. W., Joseph, S. W., ... Smith, A. C. (2007). A faculty team works to create content linkages among various courses to increase meaningful learning of targeted concepts of microbiology. *CBE Life Sciences Education, 6*(2), 155–162.

Marbach-Ad, G., McAdams, K., Benson, S., Briken, V., Cathcart, L., Chase, M., ... Smith, A. (2010). A model for using a concept inventory as a tool for students' assessment and faculty professional development. *CBE Life Science Education, 9*, 408–416.

Marbach-Ad, G., Schaefer Ziemer, K. L., Thompson, K. V., & Orgler, M. (2013). New instructor teaching experience in a research-intensive university. *Journal on Centers for Teaching and Learning, 5*, 49–90.

Marbach-Ad, G., Shaefer-Ziemer, K., Orgler, M., & Thompson, K. (2014). Science teaching beliefs and reported approaches within a research university: Perspectives from faculty, graduate students, and undergraduates. *International Journal of Teaching and Learning in Higher Education, 26*(2).

McKenna, A. F., Froyd, J., King, C. J., Litzinger, T., & Seymour, E. (2011, February 7–8). *The complexities of transforming engineering higher education.* Paper presented at the National Academy of Engineering Forum – The impact and diffusion of transformative engineering education innovations, New Orleans, LA.

McShannon, J., Hynes, P., Nirmalakhandan, N., Venkataramana, G., Ricketts, C., Ulery, A., & Steiner, R. (2006). Gaining retention and achievement for students program: A faculty development program. *Journal of Professional Issues in Engineering Education and Practice, 132*, 204–208

Nelson, K. C., Marbach-Ad, G., Thompson, K. V., Shields, P., & Fagan, W. F. (2009). MathBench biology modules: Web-based math for all biology undergraduates. *Journal of College Science Teaching, 38*, 34–39.

Penny, A. R., & Coe, R. (2004). Effectiveness of consultation on student ratings feedback: A meta-analysis. *Review of Educational Research, 72*(2), 215–253.

Quimby, B. B., McIver, K. S., Marbach-Ad, G., & Smith, A. C. (2011). Investigating how microbes respond to their environment: Bringing current research into pathogenic microbiology course. *Journal of Microbiology & Biology Education, 12*, 176–184.

Redish, E. F., Bauer, C., Carleton, K. L., Cooke, T. J., Cooper, M., Crouch, C. H., ... Zia, Z. (2014). NEXUS/Physics: An interdisciplinary repurposing of physics for biologists. *American Journal of Physics, 82*, 368–377.

Senkevitch, E., Marbach-Ad, G., Smith, A. C., & Song, S. (2011). Using primary literature to engage student learning in scientific research and writing. *Journal of Microbiology and Biology Education, 12*, 144–151.

Silverthorn, D. U., Thorn, P. M., & Svinicki, M. D. (2006). It's difficult to change the way we teach: lessons from the integrative themes in physiology curriculum module project. *Advances in Physiology Education, 30*, 204–214.

Singer, S. (2008, June 30). *Linking evidence and learning goals.* Paper presented at the National Academies of Sciences. Workshop – Evidence on promising practices in undergraduate science, technology, engineering, and mathematics (STEM) Education, workshop 1, Washington, DC.

Sirum, K., Madigan, D. L., & Klionsky, D. (2009). Enabling a culture of change: A life sciences faculty learning community promotes scientific teaching. *Journal of College Science Teaching, 38*, 38–44.

Sorcinelli, M. D. (1988). Satisfaction and concerns of new university teachers. In J. D. Kurfiss (Ed.), *To improve the academy* (pp. 121–131). Stillwater, OK: POD/New Forums Press.

Sorcinelli, M. D., Austin, A. E., Addy, P. L., & Beach, A. L. (2006). *Creating the future of faculty development: Learning from the past, understanding the present.* Bolton, MA: Anker.

Tagg, J. (2010). Teachers as students: Changing the cognitive economy through professional development. *Journal on Centers for Teaching and Learning, 2*, 7–35.

Thompson, K. V., Chmielewski, J. A., Gaines, M. S., Hrycyna, C. A., LaCourse, W. R., & Bauerle, C. (2013). Competency-based reforms of the undergraduate biology curriculum: Integrating the physical and biological sciences. *CBE-Life Sciences Education, 12*, 162–167.

Vescio, V., Ross, D., & Adams, A. (2008). A review of research on the impact of professional learning communities on teaching practice and student learning. *Teaching and Teacher Education, 24*, 80–91.

Walvoord, B. E., & Anderson, V. J. (2010). *Effective grading: A tool for learning and assessment in college* (2nd ed.). San Francisco, CA: Jossey-Bass.

Wenger, E. (1998). *Communities of practice: Learning, meaning and identity.* Cambridge, UK: Cambridge University Press.

Wieman, C., Perkins, K., & Gilbert, S. (2010). Transforming science education at large research universities: A case study in progress. *Change.*http://www.changemag.org/Archives/Back%20Issues/March-April%202010/transforming-science-full.html

Chapter 5
Preparing Graduate Students for Their Teaching Responsibilities

Teaching is hard ... but there is hope that I can systematically improve.
I think that knowing how to best teach a subject is difficult, and vital to a good learning environment. Therefore, it is very important to prepare those who will teach.

–Feedback from graduate students who took a 2-credit course
on teaching college-level biology and chemistry

Many graduate students, especially those who have just begun their graduate training, feel unprepared to teach. Professional development in teaching is important not only to improve their ability to perform as graduate teaching assistants (GTAs), but also for their future careers. Even those who do not foresee themselves in careers with an explicit teaching component can benefit from learning how to communicate complex ideas to a variety of audiences. The Teaching and Learning Center (TLC) has developed a multifaceted professional development program to address the varied needs and interest levels of GTAs. At the heart of the program is the philosophy that teaching, like research, is a scholarly activity that requires intellectual engagement. Programming includes a mandatory 6-week training course for all new GTAs and a multi-dimensional certificate program for graduate students who seek to develop greater expertise in teaching and learning. In this chapter, we provide an overview of the different components of the professional development program for GTAs, share teaching materials and assessment tools, and review data that support the effectiveness of these activities.

In Chap. 3, we demonstrated that most new faculty members have little training in teaching prior to assuming their first faculty position. Research on the experiences of graduate students indicates that much of their graduate school socialization and training focuses on their future research responsibilities, with less emphasis placed on building teaching skills (Austin, 2002; Golde & Dore, 2001; Nyquist et al., 1999). Training in teaching has immediate utility for the large majority of graduate students who serve as GTAs for at least part of their graduate education (Marbach-Ad, Shields, Kent, Higgins, & Thompson, 2010). This training also enhances graduate students' research capabilities (Feldon et al., 2011). Even those who do not plan to pursue teaching careers can benefit from training in communicating science to diverse audiences (Caserio et al., 2004). Training in teaching and communicating

© Springer International Publishing Switzerland 2015
G. Marbach-Ad et al., *A Discipline-Based Teaching and Learning Center*,
DOI 10.1007/978-3-319-01652-8_5

is particularly valuable when tailored to a disciplinary context, because disciplinary norms will strongly influence working environments and professional relationships (Austin, 2002).

In the following pages, we highlight research findings on the need for professional development for graduate students and on types of training currently offered. We then discuss how we built our program of professional development based on this research and our own needs assessment, and describe the key elements of our program.

Need for Training

The needs of GTAs change over time, so training and support must be tailored to the different stages of their development as teachers (Austin et al., 2009; Nyquist et al., 1999). GTAs should receive some formal preparation before they undertake the complex task of teaching (Prieto & Altmaier, 1994). As they develop as teachers through their initial experiences in the front of the classroom, they may desire additional opportunities for professional development to hone basic skills, such as classroom management, as well as more advanced teaching skills (Luo, Bellows, & Grady, 2000; Prieto & Altmaier, 1994). Differential training should also be offered to graduate students who are interested in pursuing teaching as a career (Austin et al., 2009).

Most existing training programs are limited in scope and focus, and do not offer the comprehensive support that graduate students need to become effective educators in the future (Gardner & Jones, 2011). Indeed, GTAs gain much of their knowledge about teaching simply through experience, often by trial and error. However, this experiential training is limited because departments generally structure graduate teaching assistantships around their own needs rather than to promote GTA development (Austin, 2002). GTAs rarely receive mentoring or feedback about their teaching (Austin, 2002; DeChenne, Enochs, & Needham, 2012), and usually are not offered opportunities to assume progressively more responsibility as they gain teaching experience (Golde & Dore, 2001).

Types of Training Currently Offered

In recent years, there has been a growing awareness of the importance of training in teaching for graduate students (Austin, 2011; Boyer Commission on Educating Undergraduates in the Research University, 1998; Gardner & Jones, 2011). Some national organizations and university graduate programs have increased their training offerings, resulting in a wide variety of training programs in teaching for graduate students. These programs differ on multiple dimensions, including sponsoring body (e.g., professional society, university, college, or department), duration, intensity, and training activities. Below, we highlight several prominent types of professional development or training.

Short Workshops for Beginning GTAs

The most common form of formal professional development for GTAs is a "one shot" workshop (Gardner & Jones, 2011). Many universities offer short workshops or orientations for new GTAs, often in the form of a multi-day workshop prior to the start of a GTA's teaching assignment. These workshops are generally university-wide, but in some cases may be offered by the department or have both university-wide and departmental components. Some universities offer a series of discrete seminars aimed at new GTAs rather than a workshop or orientation (Hollar, Carlson, & Spencer, 2000). Within a single institution, different populations of GTAs may receive different types or amounts of training. For example, it is not uncommon for international graduate students to receive additional training (Luo et al., 2000).

Longer Preparatory or Support Course

Some universities offer longer preparatory courses, which may be credit-bearing graduate courses and/or components of a professional development credential program (Addy & Blanchard, 2010). These preparatory courses are often discipline-specific (Tanner & Allen, 2006). Discipline-specific courses can center on pedagogical content knowledge and also foster the development of a community of peers with a shared interest in teaching and learning (Marbach-Ad et al., 2012). These courses generally employ active learning and evidence-based teaching approaches to model best practices that GTAs can incorporate into their own teaching (Austin et al., 2009).

Teaching as Professional Development

Professional development occurs not just through formalized training activities but also through experience gained as GTAs. Some types of teaching experience benefit graduate students more than others. For example, French and Russell (2002) found that teaching inquiry-based laboratories helped graduate students develop their ability to explain science to others. GTAs who worked collaboratively to address learning goals and improve instruction in their class or lab section, generally under the supervision of a coach or support course instructor, enhanced their teaching expertise (Alvine, Judson, Schein, & Yoshida, 2007). GTAs who worked collaboratively also engaged in more reflection on their teaching and its impact on student learning than those who worked in isolation (Dotger, 2011). Additionally, GTAs who have more instructional autonomy may have more opportunities to implement practices learned in training programs and to reinforce their training through practice (Hardré & Burris, 2012).

Being Observed

Another valuable form of professional development is having faculty members or other GTAs observe a GTA teaching in the classroom. In some cases, the observation is explicitly intended to provide feedback to the GTA as a professional development tool. For example, Bond-Robinson and Rodriques (2006) record GTAs in the classroom. They then analyze the videos, and engage GTAs in a discussion of and reflection on their teaching practices. In other cases, GTA observations occur as part of the personnel supervision structure for GTAs.

Observations can be incorporated into other professional development activities, such as mentorships or courses on teaching and learning (Marbach-Ad, Shields, et al., 2010; Prieto, Yamokoski, & Meyers, 2007; Rushin et al., 1997). In our university, observations that occurred as a part of the mandatory prep course for new GTAs were considered to be very helpful. Students noted that the observer's feedback helped them identify weaknesses in their teaching or areas to target for improvement (Marbach-Ad, Shields, et al., 2010).

Observation of Experienced Teachers

Some training programs include observations of faculty members or experienced GTAs. These observations generally occur as a component of a broader training program and are intended to complement training on best practices in teaching (Prieto et al., 2007). Observing others teach can help GTAs become reflective teachers who critically evaluate how those practices work in different settings (Schussler et al., 2008).

Certificate Programs

Some universities or departments offer a certificate program focused on teaching and learning at the post-secondary level. These certificate programs generally require completion of multiple professional development activities, including coursework, short workshops, teaching observations and evaluations, and/or the development of a teaching portfolio (Addy & Blanchard, 2010; Austin et al., 2009; Tanner & Allen, 2006). As with other professional development activities, certificate programs can benefit from a discipline-specific focus (Austin & McDaniels, 2006).

National Professional Development Networks

National networks allow individuals and institutions in diverse locations to connect around shared topics of interest and to leverage their shared resources. The NSF-funded Center for the Integration of Research, Teaching and Learning (CIRTL; www.cirtl.net) is a well-known professional development network that seeks to create model programs to prepare future STEM faculty to excel in teaching as well as in research (Austin, Connolly, & Colbeck, 2008). To achieve this goal, CIRTL promotes professional development across institutions in addition to its institution-specific programs (Austin et al., 2008; Bouwma-Gearhart, Millar, Barger, & Connolly, 2007). CIRTL provides synchronous and asynchronous online professional development activities and fosters in-person collaboration by sponsoring exchanges of individuals between member institutions (Greenler & Barnicle, 2011; Micomonaco, 2011).

A Survey of Professional Development Activities for Biology GTAs

A recent survey provides a picture of current professional development practice (Marbach-Ad, Schussler, Miller, Ferzli, & Read, 2015). The survey was conducted by the Biology Teaching Assistant Project (BioTAP), a NSF-funded initiative that connects faculty and staff who train biology GTAs in postsecondary institutions in the US and Canada. Based on the responses from 91 faculty and staff members from 81 different institutions, they found that existing practices vary widely in terms of their extent, content, and perceived effectiveness. The majority of the represented institutions indicated that they spent no more than 25 h annually on professional development in teaching for biology GTAs, with about 40 % spending ten hours or less per year. Most institutions offered pre-semester orientations, either for individual courses, the department, or the institution. The most commonly covered topics included classroom management, course content, teaching policies, and teaching techniques. Survey responses did not demonstrate a strong relationship between the amount of time dedicated to professional development and the perceived effectiveness of professional development. Despite this weak relationship, more than half of the respondents reported that professional development offerings for GTAs had a positive impact on their teaching effectiveness.

For more information on BioTAP, see www.bio.utk.edu/biotap/.

Professional Development for GTAs in Our University

Our university, like most others, offers a variety of orientation and professional development programs for graduate students in general and GTAs in particular. Each semester, all incoming graduate students are invited to attend a one-day orientation hosted by the campus-wide Center for Teaching Excellence (CTE). This orientation includes breakout sessions on various topics of relevance to graduate students. Within the College, each department also offers a two-day orientation to the graduate program that includes a few sessions related to the roles and responsibilities of GTAs. Additionally, the CTE and other campus groups host workshops throughout the year specific to graduate students. GTAs are also referred to additional training provided by professional societies and networks such as CIRTL.

Our discipline-based Teaching and Learning Center, which serves the biology and chemistry departments, offers professional development in teaching that is rooted in the principles of pedagogical content knowledge (PCK; see Chap. 1 for more information about PCK and its components). To tailor our professional development program to the specific needs of the biology and chemistry graduate students in our College, we conducted a needs assessment.

Needs Assessment

In order to tailor our training and support programs to the College population, we conducted a need assessment that included a survey of graduate students and a workshop for both faculty and graduate students on the role of the GTA in undergraduate education.

Key Findings from the Survey

We surveyed all graduate students about their experiences as GTAs, their preparation for this role, and the challenges that they faced. We also sought their input on how we could meet their professional development and training needs related to their GTA responsibilities. Fifty-two individuals, representing approximately 30 % of all biology and chemistry GTAs, responded to the survey.

Our GTAs were assigned a variety of teaching responsibilities. Almost all respondents indicated that they had supervised or taught a lab course as a GTA. About half had led discussion or recitation sections of large lecture courses.

We asked survey respondents to rate various sources of professional development in terms of their helpfulness in preparing them for their GTA responsibilities, using a scale of one (Not at all helpful) to five (Extremely helpful). As Fig. 5.1 shows, survey responses indicated that the most helpful sources of professional development (in order of helpfulness) were lab/recitation coordinators for the

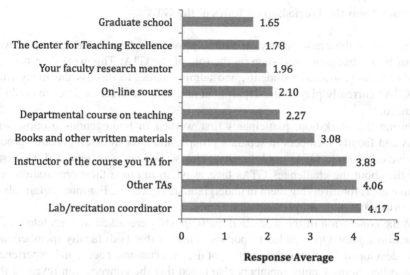

Fig. 5.1 The helpfulness of different resources for GTAs, on a scale of 1 (Not at all helpful) to 5 (Extremely helpful)

course, other GTAs, and the instructor of the course in which they served as a GTA. Optional departmental teaching preparatory courses on teaching received an intermediate rating, although only 27 of 52 respondents reported having taken this course. Campus-wide sources, such as programs sponsored by the graduate school and campus-wide teaching center, were rated on the low end of the scale.

The survey included an open-ended item that asked respondents for suggestions about how the College or department could better prepare GTAs for their teaching responsibilities. Of the 42 responses, 14 respondents indicated that they would like additional professional development in teaching and learning. They offered various suggestions for what form this professional development should take, such as additional support from lab coordinators and/or workshops or courses on teaching. Some responses outlined specific topics to be covered in the workshops or courses:

> I would love to see a formal teaching course including how to prepare exams and homework assignments, how to translate from a textbook to a lesson plan, etc.
> Workshops to teach how we select topics to be incorporated in the class material, how to design lab courses and experiments for the labs, how to incorporate new technologies to enhance the teaching experience in the classrooms.
> Training in UMD technologies. Training in time management.

Some respondents suggested that their professional development should include enhanced responsibilities as a GTA, such as involvement in course design and delivery of lectures. Others suggested putting more consideration into graduate students' workload and better tailoring course assignments to GTA research interests and background. Responses also suggested that discipline-specific training, and training related to common GTA assignments such as leading lab sessions, would be helpful.

Feedback from the Workshop on Roles of the GTA

The results of the graduate student survey provided a foundation for the conversations in a subsequent workshop on the role of the GTA. This workshop brought together faculty, graduate students, and administrators to discuss the many roles that GTAs currently play in the department and how the GTA experience could be enhanced.

During the workshop, participants first worked in homogenous groups, with GTAs and faculty members in separate groups, and later in heterogeneous groups that included both GTAs and faculty. The workshop was intended to foster dialogue about the challenges GTAs face, ways to improve their preparation, and creative ideas for involving them in undergraduate education. For more details about the workshop, see the Implementation Guide.

At the conclusion of the workshop, participants were asked to complete a brief evaluation survey. Open-ended responses indicated that both faculty members and GTAs developed a better understanding of the expectations, needs, and experiences of the other. Some faculty members also noted that the conversation revealed that some courses give no direction to GTAs. Survey respondents also recommended that some of the issues brought up in the workshop should be addressed through additional professional development for GTAs.

A direct outcome of this workshop was the revision of the optional prep courses as mandatory, team-taught preparatory courses with one offering for new biology GTAs and another for new chemistry GTAs (described below). We also created tip sheets on classroom management, grading, communicating with students, expectations and evaluations of GTAs, and recommended teaching strategies that are available as an online resource (www.cmns-tlc.umd.edu/resourcesforGTAs).

Our Professional Development Programs for GTAs

Based on our needs assessment and a literature review of the different models for professional development (see Types of Programs Currently Offered, above), we developed a program that offers variety of activities targeted at different levels of GTA interest and expertise in teaching. Our activities differ in their intended audience, their duration, and the degree to which participation is compulsory. The program includes four key components:

- Seminars and workshops
- Mandatory, six-week courses for new GTAs
- An optional, two-credit course
- A certificate program in teaching and learning

Seminars and Workshops

Graduate students are encouraged to attend the TLC's regularly scheduled seminars and workshops, which are generally targeted at a broad College audience, including postdocs, faculty members, and administrators, as well as graduate students. These seminars and workshops cover a broad range of topics related to all areas of PCK, as described in more detail in Chap. 2 of this volume.

In addition to these seminars and workshops, each semester we host a Visiting Teacher/Scholar (also described in Chap. 2). During these visits, we hold a lunch in which interested graduate students can engage in open, informal conversation with these prominent scholars who have contributed to the growth of science education research. In these conversations, the participating graduate students generally inquire about broad career goals, such as how to become a strong candidate for a postdoctoral or faculty position and how to balance competing demands on their time. They also ask about the visitor's specific experiences and science education research. We generally limit the lunch to around ten participants to maintain a more intimate conversation. If the TLC knows that graduate students are involved in projects related to the visitor's science education research, we make a specific effort to encourage them to attend the lunch or arrange for an individual meeting between them and the Visiting Teacher/Scholar. By making graduate students integral participants in these visits, we seek to reinforce their important instructional role within the university and help them develop their identities as educators.

Mandatory Six-Week Courses for New GTAs

One of the themes that emerged from our needs assessment was a desire on the part of GTAs for more training in preparation for the challenges of their first teaching assignment. As a result, we developed two six-week preparatory courses that are mandatory for all new GTAs in the biology and chemistry departments. The discipline-specific courses are designed to complement the initial orientations provided by the university and/or departments by helping new GTAs develop pedagogical content knowledge.

> *I never taught before, so this course taught me a lot.*
>
> *The course provided a nice support system... This class helped provide necessary information for grading and what to expect from my students and how other TAs handled situations.*
>
> –Student feedback from the end-of-course evaluation survey

Goals of the Courses

Preparatory courses for graduate teaching assistants should be designed to accomplish several goals: building a community, modeling best practices in teaching, and providing a venue for reflection on teaching.

Build a Community for New GTAs that Socializes Them into the Department

All of the students in the preparatory courses are new to their graduate program. Most are also new to the university and, in our university, many are international students who are new to the country. We believe that these courses help graduate students build connections with more experienced GTAs and faculty members within their department, and assist in their socialization into the departmental culture. In the case of international students, the courses also serve to introduce them to American university and cultural norms. To achieve these goals, many faculty members and experienced GTAs participate in delivering the courses. One way we incorporate their participation is through a team-teaching approach in which a science education specialist collaborates with one or more faculty members from the department to lead the course. TLC staff and DBERs from the respective department are also involved in planning and delivering the course. Before the courses begin, the department chairs request that faculty members and experienced graduate students sign up to participate in a course session appropriate to their interests and expertise in teaching. Each course session involves two to three guests, who offer different perspectives based on their differing roles (e.g., lab coordinators, senior faculty, lecturers, discussion session leaders). The chairs of the departments play important roles in the delivery of the courses, serving as co-instructors or offering a presentation at the first class meeting. This participation demonstrates the value that the department places on teaching and helps graduate students build connections to departmental leadership.

> *I learned some good tips and formed good relationships with other TAs.*
>
> –Student feedback from the end-of-course evaluation survey

Model Best Practices in Teaching

In addition to community building, these courses seek to highlight effective instructional approaches. We incorporate evidence-based teaching practices, including work in small groups, problem-solving activities, simulations, role play, think-pair-share, multimedia presentations, and a general focus on active rather than passive learning techniques. Through these practices we explicitly and implicitly promote the teaching approaches that we want graduate students to adopt, and also model effective practices that may be unfamiliar to some students in the class.

Provide GTAs with a Venue for Reflecting on Their Teaching

We gave much consideration to the timing of the course. We opted to offer the course during graduate students' first semester on campus, which generally coincides with their first teaching assignment. We want them to be teaching and enrolled in the course concurrently, rather than taking the course before they start teaching, so that they have a more realistic understanding of the associated responsibilities and challenges. The course offers them an opportunity for timely reflection on their teaching and allows them to seek solutions to problems that they encounter in assuming this new role. Additionally, course discussions are richer because students can share their recent experiences from the classroom.

Sequence of Topics

The course is intended to cover a broad range of topics and prepare GTAs for their initial experiences leading classes, particularly lab sections. Course structures and content vary across departments and over time, based on the graduate student audience and the staff leading the course. Table 5.1 presents a suggested sequence of topics mapped to the five PCK components (see Chap. 1 for more details on each component). The course covers four of the five PCK components; it does not cover knowledge of science curriculum, because few GTAs are involved in decisions related to curriculum. In the Implementation Guide, we provide a detailed description of the course, including the topics covered in each course session and the format in which those topics are covered. We also offer recommended readings for each session.

One of the most successful components of the course has been the "stories from the field" offered in each class session by experienced GTAs and course instructors. Class members are also encouraged to share their own experiences during these discussions. These stories provide new GTAs with a better understanding of how to address the difficult situations that inevitably arise in the classroom.

It was great to hear first-hand stories of unusual situations that may arise and how people have handled those.

Hearing about other TA (and instructor) experiences helped me deal with my own issues over the course of the semester.

'War' stories brought up problems/solutions that I would otherwise not think of.

–Student feedback from the end-of-course evaluation survey

Table 5.1 Sequence of topics and corresponding PCK components for a 6-week teaching preparatory course

Topics	PCK component(s)
1. Introduction to the course Overview of the course The role of the GTA in undergraduate education	5. Orientation to science teaching
2. Student-GTA communication Tips for communicating effectively with students Handling difficult situations in communication	3. Instructional strategies 5. Orientation to science teaching
3. Assessing student performance Grading exams and lab reports Maintaining consistency in assessment Writing assessment items	1. Student understanding of science 4. Assessment of student learning
4. Engaging students through effective teaching Tips for effective teaching Handling difficult situations in the class	3. Instructional strategies
5. Evaluation and expectations of the GTA Personal strengths and weaknesses Interpreting student evaluations of GTAs	3. Instructional strategies 5. Orientation to science teaching
6. Special topics based on GTA interests Modeling presentations of introductions to laboratory exercises Alternative scientific careers Interactive game about student confidence	3. Instructional strategies 5. Orientation to science teaching

Optional Two-Credit Course

In response to a need for more intensive professional development in science education, particularly for students who plan to pursue a career involving science education, we developed a two-credit graduate course on teaching and learning in the biological and chemical sciences. This course is open to all graduate students in the biology and chemistry departments, and is taken by some students to fulfill requirements for specific graduate fellowships (e.g., those funded by the University of Maryland's Graduate Assistance in Areas of National Need [GAANN] grant). The course is also a required component of our certificate program in teaching and learning (described in more detail later in this chapter).

Prior to the development of this course, some students from our College took a university-wide graduate course on teaching and learning. As we saw greater interest in more training in science education, the TLC and College/departmental leadership developed a course that targets the specific needs of biology and chemistry graduate students. In addition to providing more advanced professional development in the theoretical foundations of scientific learning and teaching methodologies, our discipline-specific course fosters a community with shared interests and backgrounds in science education.

Table 5.2 Sequence of topics and corresponding PCK components for a two-credit course on science teaching and learning

Class sessions and topic covered	PCK component(s)
1. Introduction to the course	5. Orientation to science teaching
2. Overview of science education	3. Instructional strategies 5. Orientation to science teaching
3. Student diversity in learning styles and multiple intelligences	1. Student understanding of science
4–5. Theories in science education	Varies based on theories covered
6. Active learning and evidence-based teaching approaches	3. Instructional strategies
7. Assessment, rubrics, and grading	4. Assessment of student learning
8. Students' alternative conceptions	1. Student understanding of science 3. Instructional strategies
9. Using visual representations: concept maps and models	2. Science curriculum 3. Instructional strategies 4. Assessment of student learning
10. Using technology in science education	3. Instructional strategies
11. Observation of an undergraduate science class	All PCK components
12–14. Presentations of teachable units	All PCK components

Sequence of Topics

The course covers a broad range of topics that build on the introduction to effective practices presented in the mandatory course, and also includes an introduction to salient theories and research in science education. Table 5.2 outlines the topics covered in the course and their corresponding PCK component(s). The course is taught in a seminar style with student-guided discussions based on weekly readings. Class activities employ a variety of active learning approaches that model best practices in teaching and learning. Students submit a series of short written assignments throughout the semester. The culminating project of the course is the development of teachable units, which are demonstrated during the final class meeting. These course assignments are intended to foster students' engagement with theories on teaching and learning and reflection on their own practices as educators. More details about how these topics are covered in the course, and the associated homework assignments and activities, are provided in the Implementation Guide.

I liked the open-ended discussions in class because you gained an appreciation for all of the different viewpoints that are available and learned different styles.

[I liked the] free-flowing yet goal-oriented discussions.

(continued)

> *The way the course was set up—laid back, roundtable discussion, pressure-free—made it easier to contribute.*
>
>   –Student feedback from the end-of-course evaluation survey

Certificate Program in Teaching and Learning

The campus Center for Teaching Excellence (CTE) at our university has long offered a graduate certificate program in teaching and learning. This program provides professional development based on the philosophy that teaching is a scholarly activity requiring intellectual engagement and conversation. The program is open to graduate students from all disciplines who have taught at least one semester as a GTA.

As a part of our discipline-based program for professional development, we collaborated with the CTE to offer a certificate program specific to biology and chemistry graduate students. This discipline-specific program offers participants the comprehensive professional development of the university-wide program, with the added benefit of instilling PCK in all aspects of the program. Our program focuses on approaches and issues that are of particular relevance to educators in STEM fields, such as scientific teaching and addressing alternative conceptions of science. We see the program as an opportunity to enhance graduate student preparation for academic careers, as well as other careers that involve communicating science content.

The program involves multiple components that participants can complete at their own pace and in any order. The required program components are shown in Fig. 5.2 and described below. We supplement these descriptions with participant feedback from program evaluation interviews (Marbach-Ad, Katz, & Thompson, in press).

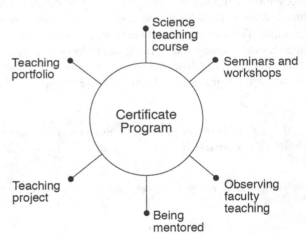

Fig. 5.2 The six required components of the graduate certificate in our university teaching and learning program

Science Teaching Course

To earn the certificate, students must successfully complete a graduate course on university teaching. We recommend that participants take the Biological and Chemical Teaching and Learning in Higher Education course offered by the TLC, described above. Students who are unable to take the disciplinary course may substitute the university-wide course sponsored by the CTE.

> *Having [the two-credit] class opened my eyes to a lot of teaching techniques ... There's a lot for undergraduate science education, for help, for support, and different journals to read. Learning more about learning theory and the psychology behind that was really helpful.*

Workshops

The certificate requires attending at least seven workshops on teaching and learning, including at least one focused on the use of technology in instruction. Students can choose from among a variety of offerings to fulfill this requirement, including workshops sponsored by the campus CTE, the campus Office of Learning Technologies, and the TLC (see Chap. 2).

> *There were constantly useful workshops and a lot of really interesting guest speakers and chances to interact with them after.*

Observations of Experienced Instructors

Participants observe classes taught by instructors noted for their exemplary teaching. After the observation session, participants write a reflection on what they learned. Participants are encouraged to observe at least three different instructors to see a variety of best teaching practices and observe how students respond to different teaching styles.

Being Mentored

Participants choose a faculty member in their department or the College to serve as a mentor in teaching. Mentors observe their mentee's teaching at least twice. Following the first observation, mentors provide verbal and written feedback to the mentee to help them improve their teaching. We strongly suggest that the mentee videotape the observed class session and write an analysis of the class to compare with that of their mentor.

Teaching Projects

Participants must complete a science teaching project that demonstrates their understanding of the field and contributes to the improvement of instruction within the department, college, or university. The TLC director, in consultation with other faculty members, supervises participants' teaching projects. Projects take many different forms, such as developing teaching tips for other GTAs, conducting a workshop on teaching and learning, or evaluating an innovative course or teaching practice. Additionally, some participants have expanded their teachable units that were initially created as the final project in the two-credit course. For more complex projects, participants are encouraged to work in pairs. Some teaching projects have been disseminated through conference presentations and peer-reviewed publications (Lombardi, Hicks, Thompson, & Marbach-Ad, 2014; Sinnott, Marbach-Ad, Orgler, & Thompson, 2011).

> *The teaching project... was actually a lot of fun. We did an experiment to test the effectiveness of a few different approaches to teaching. I worked with another person from the UTLP on that, and we got to work with undergrad students. We did an actual controlled test for these different teaching approaches... We ended up presenting that work at a conference and got that paper published.*

Teaching Portfolio

Participants develop a teaching portfolio that documents their expertise, experience in, and ideas about teaching. Teaching portfolios should include at least the following:

- A statement of teaching philosophy
- A description of teaching responsibilities at the university
- A reflective summary of student evaluations of teaching, covering at least two semesters of classes
- A statement about efforts to improve teaching and to contribute to public discussions of teaching and learning

To assist students in assembling these materials, the campus CTE offers a three-day retreat on teaching portfolio development. Participants prepare drafts of their materials in preparation for the retreat, then receive critiques and have the opportunity to fine-tune their materials during the retreat. The completion of the teaching portfolio is an excellent opportunity for students to prepare statements of teaching philosophy that are often required when applying for faculty positions.

I think the teaching portfolio was really important. I could never have done that on my own.

Graduate students who successfully complete all of the required program components receive a University Teaching and Learning Program (UTLP) certificate and UTLP notation on their graduate transcript.

[The UTLP] diversifies you a bit. It helps you stand out, and it helps prepare more your science education side. A lot of students don't really have that background. I think it is mindboggling that to teach middle school, you student teach for a semester and you have about 25 credits of education [coursework], but to teach college, nothing. A lot of students have one seminar and that's it.

We networked and we met other people who liked doing the things that we did, which wasn't necessarily all that common in our school's doctoral program. You have to TA to get your funding; it's not necessarily something you like to do. So it was nice to meet other students who had similar interests.

Conclusion

Professional development programs in teaching for GTAs are usually insufficient (Boyer Commission on Educating Undergraduates in the Research University, 1998; Gardner & Jones, 2011; Golde & Dore, 2001). Although many graduate students serve as teaching assistants in undergraduate courses, they often receive a strong message to focus on their research and not devote much time to teaching (Brownell & Tanner, 2012; Nyquist et al., 1999). In accord with this prioritization, departments and colleges usually do not offer substantial professional development programs to their GTAs (Austin, 2002; Austin et al., 2009; Luft, Kurdziel, Roehrig, & Turner, 2004). This lack of training is especially disappointing since many GTAs report that they are interested in developing teaching proficiency (Tanner & Allen, 2006) and see themselves continuing to careers that include teaching (Sauermann & Roach, 2012). Teaching and learning centers can play an important role in changing this institutional landscape by facilitating changes that improve GTAs' teaching effectiveness and enhance their preparedness for their future careers (Brownell & Tanner, 2012; Kendall et al., 2013).

Sensitivity to the diversity of professional aspirations among GTAs led us to develop TLC programming that includes four primary components: seminars and workshops, a mandatory six-week course for new GTAs, an optional two-credit course, and a certificate program in teaching and learning. We have found that these four components allow graduate students the flexibility to select options that correspond to their level of experience, interests, and career goals. The design of the program is based on the science education literature as well as the specific needs and context of our own university.

While we recommend a comprehensive program of professional development activities, the specific content and structure of that program can and should vary across contexts. An initial needs assessment and ongoing evaluation are important tools for tailoring professional development to suit the institution and graduate student populations served.

Implementation Guide

The Implementation Guide provides an overview of the content and structure of a workshop on the role of the GTA and our two course offerings for GTAs. This Implementation Guide is intended to facilitate the adoption and/or adaptation of these activities by other institutions.

Workshop: The Role of the GTA

In this workshop, faculty members and graduate students collaborated first in homogenous and then in heterogonous small groups to discuss the role of the GTA and how to enhance the professional development and support offered to GTAs. The 90-min workshop included the following components:

- Summary of GTA needs assessment survey results (10 min)
- Small group conversations on GTA roles and challenges (25 min)
- Small group conversations on enhancing GTA professional development (25 min)
- Panel on faculty and GTA experiences with innovative course redesign (20 min)
- Question and answer session (10 min)

Summary of GTA Needs Assessment Survey Results

At the start of the workshop, the facilitator reviewed the results from a needs assessment survey that had been administered to GTAs in the weeks preceding the workshop. The survey results provided an overview of the types of responsibilities that GTAs had in our College, the resources they viewed as most influential in preparing them for teaching, and their suggestions for additional professional development programming.

Conversations on GTA Roles and Challenges

Participants then divided into homogenous groups of either faculty members or graduate students, with three to five participants per group. Groups of faculty members and graduate students were given slightly different sets of discussion questions.

Questions for GTAs

1. Do you think it is important for graduate students to be involved in undergraduate education?
2. Do you think that when you started your teaching responsibilities you were qualified to do so

   (a) in terms of content knowledge?
   (b) in terms of knowing how to teach?

3. What challenges do you face as a GTA?
4. What are your concerns regarding your teaching responsibilities as a GTA?

Questions for Faculty

1. Do you think it is important for graduate students to be involved in undergraduate education?
2. Do you think that GTAs are usually qualified for their teaching responsibilities

   (a) in terms of content knowledge?
   (b) in terms of knowing how to teach?

3. What are the challenges faced by GTAs?
4. What are your concerns regarding the teaching responsibilities of the GTAs?

Following these small fes of each small group discussion group discussions, the entire group came back together and shared their responses.

Conversations on Enhancing GTA Professional Development

Participants again divided into groups of three to five, this time mixing graduate students and faculty within each group. Groups discussed how GTAs could enhance the undergraduate classroom experience. They also talked about ways to better support GTAs so that they are prepared to and capable of improving undergraduate education. To guide these conversations, each group responded to the following prompts:

1. Share creative ideas to involve GTAs in teaching.
2. Share ideas for ways to enhance professional development and support for GTAs.

Again, the group reconvened to hear synopses of each small group discussion.

Experiences with Innovative Course Redesigns

To demonstrate the potential of involving GTAs in expanded teaching roles, we selected a recent course redesign in which graduate students were deeply involved in all stages of developing and implementing the new course. The participating faculty members and GTAs shared their experiences redesigning the discussion session for a large-enrollment biology course. The redesign occurred in response to low student attendance at the course's optional discussion sessions, as well as student feedback indicating that the format of the sessions was not ideal. To make the discussion sessions more valuable to students, the course instructor worked with the TLC, a discipline-based education researcher, and two GTAs to implement a Just-in-Time Teaching (JiTT) approach (Novak, Patterson, Gavrin, & Christian, 1999). For each discussion session, the GTAs developed questions that students were required to answer and submit prior to the session. The GTAs then reviewed student responses, identified common misunderstandings, and addressed these problematic concepts in a new format that engaged students in small group discussions. The GTAs shared how they benefitted from their enhanced role in developing and guiding the discussion sessions.

Mandatory Six-Week Teaching Preparatory Course

The mandatory teaching preparatory course is structured to prepare GTAs for their instructional roles. Below, we provide an overview of the course logistics, then describe course sessions and related activities. Each session generally involves the participation of the course co-instructors (a science education specialist and a faculty member from the department), as well as guest faculty members and a few experienced GTAs.

In designing a professional development program that would be required for all new GTAs, we had to balance training needs with the many competing demands on graduate student time. We found that the six-week duration was manageable for GTAs while still offering sufficient time to cover key topics that can help them grow in their instructional roles. It can be difficult to find a time in which all new GTAs are available for the course, particularly in a large department with many new incoming graduate students. In our university, we have found that an evening course is most convenient. We have established 6:30 to 8:30 pm on Tuesdays as the standard time for the courses, and work with departmental leadership to ensure that new graduate students are not assigned other responsibilities in this timeframe. If a course coincides with a meal time and funding is available, providing food creates a collegial atmosphere. It further demonstrates that the department values teaching and is willing to allocate resources to improving instruction.

Session 1: Introduction and Overview of the Role of the GTA in Undergraduate Education

Introduction

At the beginning of the first session, we ask students to complete a short questionnaire probing their past experience and expectations regarding their teaching. Students respond to the following questions:

- Have you had prior experience as a Teaching Assistant? If so, please describe the course (title) and experience.
- What are you most excited about in undertaking your role as a Graduate Teaching Assistant?
- What are you most afraid of in undertaking your role as a Graduate Teaching Assistant?
- How will you manage your teaching to ensure that course sessions are productive?

After students have completed the survey, the course co-instructors provide an overview of the syllabus. Students introduce themselves and describe their background and career aspirations. We also ask students to wear nametags so that they can start getting to know their peers.

The Role of the GTA

We provide an overview of the role of the GTA in course delivery, the importance of GTAs within the department, and performance expectations for GTAs. Because GTAs are all new to the university, this is an opportune time to discuss norms, policies, and processes. We sometimes also distribute documents related to the honor code and university policies on research ethics. We recommend that the department chair deliver this introductory presentation because the chair's leadership position lends credibility and authority, in addition to emphasizing the value that the department places on the GTAs' contributions to undergraduate instruction.

Next, a panel of three to five experienced GTAs describes their responsibilities as teaching assistants, discusses challenges or difficulties that they have encountered, and offers recommendations for success in teaching. In addition to the panel, we also invite lab coordinators to share their perspective on the roles and responsibilities of GTAs in laboratory courses. The course co-instructors and all presenters repeatedly emphasize that they are available to support GTAs with any concerns they may encounter in the future.

Following these presentations, we randomly divide all class participants into small groups, with one faculty member or experienced GTA in each group. The small groups, and later the entire class, discuss the following questions:

- What expectations do you have regarding your GTA responsibilities? Specifically, what are your responsibilities, and how do you expect to fulfill them?

- What were the characteristics of your best teachers? What made them particular effective at teaching you?
- What do you expect of the students in your classes? What do you expect regarding their background knowledge of [discipline]? What do you expect regarding their motivation to study [discipline]?

Recommended Readings/Resources

These recommended readings and resources are concise, accessible articles that were selected to provide course instructors with background information in preparation for class sessions. They are also recommended to students who want to enrich their knowledge about teaching and learning in the sciences. The papers can also be assigned to students before a given class session to serve as the basis for an in-class discussion.

What Defines Effective Chemistry Laboratory Instruction? Teaching Assistant And Student Perspectives (Herrington & Nakhleh, 2003).

Graduate Students' Teaching Experiences Improve Their Methodological Research Skills (Feldon et al., 2011).

Session 2: Student-GTA Communication

Stories from the Classroom

A panel of four to six faculty members and experienced GTAs share stories about their own experiences with very successful or very problematic communications with students. The whole group discusses the stories and suggests alternative ways of dealing with the situations. New GTAs are invited to share their own stories and to seek feedback on how to communicate more effectively with their students.

Student-GTA Communication Scenarios

Participants split into small groups, with one faculty member or experienced GTA in each group. Each group discusses possible responses to various student-GTA communication scenarios.

Sample Scenarios

Scenario 1—The student that continually asks questions
A student in your lab/discussion continually asks you questions that cause your discussion section or lab instruction to be disrupted. Also, what if you have a student that persists on following you around even after office hours with questions? What do you do or say to this student?

(continued)

Scenario 2—The non-participating student
A student in your discussion/lab either does not want to participate in the discussion or with his/her group on an experiment that the class is working on. In the case of a laboratory experiment, the student sits there and either listens to music or plays around with his/her computer instead of helping their partner who is doing all of the lab work. What do you do or say to this student?

For more scenarios, see Marbach-Ad et al. (2012).

Recommended Readings/Resources

All Students Are Not Created Equal: Learning Styles In The Chemistry Classroom (Bretz, 2005).
    Reaching The Second Tier: Learning And Teaching Styles In College Science Education (Felder, 1993).

Session 3: Assessing Student Performance

Stories from the Classroom

A panel of faculty members and experienced GTAs share stories related to grading and student assessment in lab and discussion sessions. The conversation then expands to encompass the whole class.

Do's and Don'ts for Grading Assessments

Lab coordinators provide an overview of grading policies and procedures. They also share tips for grading lab reports and exams, and discuss common pitfalls in grading and communicating with students about grades. The presenters also discuss the need to be aware of and sensitive to individuals with disabilities, and share university policies related to special accommodations for these students.

Grading Exercise

The instructor distributes examples of actual undergraduate student responses on exams and lab reports, along with answer keys or rubrics for scoring these responses. The example items are representative of the types of items appropriate to the discipline. For example, the chemistry GTA course includes mathematical calculations, chemical equations, and constructed response items in which students briefly explain a concept, discuss lab procedures, or justify their answer to a question. Class participants first grade the student responses individually using the answer key or rubric. Participants then work in small groups to review their grading of the student responses, discuss any differences, and reach a consensus grade for each item. The whole class then reconvenes to share their consensus grades and discuss differences in how the items were graded. This exercise demonstrates the challenges in maintaining consistency of grading among multiple graders, even when a common answer key or grading rubric is used. The discussion concludes with ideas for minimizing inconsistencies in grading student responses.

Writing Test and Quiz Items Exercise

Class participants are introduced to Bloom's taxonomy (Bloom, 1984) and asked to create at least three test or quiz items, each representing a different level of Bloom's taxonomy (knowledge, understanding, application, analysis, synthesis, and evaluation). The group then reviews and critiques all of the items in terms of their quality, clarity, and cognitive level, and revises a subset of these items.

This exercise can be done in various ways. Rather than doing the entire exercise during class, the item writing can be assigned as homework so that the class meeting time can focus on reviewing, critiquing, and discussing the items. Alternatively, class participants may create items and post them to a course website for peer review prior to the class session.

Recommended Readings/Resources

Taxonomy of Educational Objectives: Handbook 1: Cognitive Domain (Bloom, 1984).

Rote Versus Meaningful Learning. Theory Into Practice (Mayer, 2002).

Bridging The Educational Research-Teaching Practice Gap. Conceptual Understanding, Part 2: Assessing And Developing Student Knowledge (Schonborn & Anderson, 2008).

Grading using rubrics was very helpful when I had to later use a rubric to grade lab reports.

I had never used a rubric before, so it was a good introduction.

–Student feedback from the end-of-course evaluation survey

Session 4: Engaging Students Through Effective Teaching

Five-Minute University

We open the session by showing the YouTube clip "Five-Minute University" in which the fictional character Father Guido Sarducci suggests a university program that takes only five minute to achieve the same long-term learning outcomes as a traditional university program. This satire offers an entertaining portrayal of a key issue in undergraduate education: many university programs or courses do not teach in a way that promotes lasting student learning.

Teaching Tips and Stories from the Classroom

This section of the workshop consists of a panel of faculty and experienced GTAs, who recount their stories from the classroom within the context of topics covered in *McKeachie's Teaching Tips* (McKeachie & Svinicki, 2014). They present the key ideas from four chapters of the book and then provide relevant examples from their own experiences. The chapters cover the following topics:

- Facilitating Discussion: Posing Problems, Listening, Questioning (Chapter 5, pp. 38–57)
- Different Students, Different Challenges (Chapter 13, pp. 172–188)
- Laboratory Instruction: Ensuring An Active Learning Experience (Chapter 19, pp. 277–288)
- Teaching Culturally Diverse Students (Chapter 12, pp. 151–171)

These four chapters offer different ways of engaging students through active learning and addressing the typical challenges that GTAs are likely to face when doing so.

Teacher-Student Interaction Scenarios

Faculty members and experienced GTAs work in pairs or small groups to act out skits modeling common teacher-student interaction scenarios. Each group prepares their five-minute skit in advance, with one person playing the role of a teacher and another playing the role of a student. Following each skit, class participants discuss the scenario and suggest how the teacher could have handled the situation more effectively.

Sample Scenarios

Scenario 1: Pushy student and patient teacher. In this scenario, the student is trying to find out from the teacher what the best possible interpretation of the

(continued)

experiment should be rather than interpreting the results by himself/herself. In other words, the student seeks to be provided with the answer to a lab question that students in the class are expected to answer independently.

Scenario 2: Multiple students competing for instructor's attention. In this scenario, three or four students all seek the sustained attention of the lab instructor, who must determine how to balance these competing demands. The students make different types of requests, and the instructor must prioritize response order based on the urgency of student need (e.g., a student unsure of the next step in a lab procedure vs. a student with a spilled chemical on his/her hands).

For more scenarios, see Marbach-Ad et al. (2012).

Recommended Readings/Resources

Why Not Try a Scientific Approach to Science Education (Wieman, 2007).
  Scientific Teaching (Handelsman et al., 2004).
  YouTube video clip: Confessions of a Converted Lecturer: Eric Mazur (www. youtube.com/watch?v=WwslBPj8GgI).

> *I learned a lot of teaching techniques ... This course helped me streamline my lectures and be more organized.*
>
> *This course did give me ideas for different teaching techniques. I could try to find something that could work for my students.*
>
>   –Student feedback from the end-of-course evaluation survey

Session 5: Evaluation and Expectations of the GTA

Personal Strengths and Weaknesses

At the time of this workshop, approximately five weeks into their teaching assignments, GTAs have begun to develop an awareness of their strengths and weaknesses as instructors. To encourage them to reflect on their teaching experience, we ask all class participants to share one personal strength and one weakness. The class discusses these strengths and weaknesses, and suggests ways to overcome difficulties faced in the classroom.

Student Evaluations of GTAs

At our university, students are asked to complete evaluations of all course instructors, including GTAs, at the end of each semester. Course instructors are provided with a detailed summary that includes aggregate ratings of the instructor's effectiveness as well as anonymous, open-ended comments.

Instructors can learn a lot about their teaching strengths and weaknesses from these student evaluations. We provide an overview of the components of the student evaluation and how results can be interpreted. We encourage GTAs to review their evaluations and pay particular attention to the open-ended comments. We also encourage GTAs to seek feedback from their students prior to the end of the semester by giving a midterm evaluation survey. This can be done using the anonymous quizzing function of the online course management system or it can be done informally, such as through an in-class paper survey. If midterm evaluations reveal problems or weaknesses, GTAs can invite a faculty member, representative from the TLC, or other GTAs to observe their teaching and offer constructive criticism. The course instructors provide tips for interpreting student evaluations:

- Not all of the items in the evaluation survey address components of the course that GTAs can control. Items that are generally beyond the control of GTAs include course curriculum and assessment.
- Some student responses may be strongly worded and reflect frustrations related to a low grade, stress at the end of the semester, or components of the course unrelated to the GTA's instruction. Instructors should try to identify substantive, constructive critiques of their teaching and can ignore other comments. Instructors should also look for patterns in student responses, and focus more attention to responses that share common themes while paying less attention to outliers.
- Scores on different evaluation items tend to be highly correlated, reflecting the students' general satisfaction with the course. A particular item or subset of items that is rated substantially lower than the rest of the items may indicate a teaching problem that needs to be addressed.
- Different populations may respond differently. For example, we have found that a night course aimed at nontraditional, adult learners tends to receive lower evaluation scores than a daytime course taught to traditional students by the same instructor.

Class participants then separate into small groups, with a faculty member or experienced GTA joining each group. Each group is given examples of student evaluations from courses in their department, with the instructors' names masked. The groups discuss key points from the evaluations and suggest ways to overcome any apparent teaching deficiencies.

Session 6: Special Topics Based on GTA Interests

We generally leave the last session of the course open and determine the specific topic(s) to be covered based on student requests, queries, and interests. Below we outline three topics that our past class participants have found to be particularly valuable. These topics can be offered individually or in combination.

Modeling Presentations of Introductions to Laboratory Exercises

This modeling activity is intended to help new GTAs hone the presentation skills necessary for introducing students to the concepts and techniques needed for a particular laboratory exercise. Many of our GTAs lead laboratory courses, and each lab session typically begins with a short presentation that provides an overview of the lab objectives and protocols. In addition to preparing the students to undertake the laboratory, these introductions connect the lab to students' existing knowledge and course content.

In the modeling activity, experienced GTAs simulate effective and ineffective presentation techniques. Effective techniques include using an enthusiastic tone and frequently pausing the presentation to engage students in questions and responses. Ineffective techniques can include using a monotonic voice, avoiding eye contact, speaking too quickly, implying that the lab is boring, presenting instructions in a disorganized fashion, not defining new terms, and not offering students an opportunity to participate or ask questions. After each model presentation, the class provides a critique highlighting its strengths and weaknesses. The group then engages in a discussion of alternative presentation styles that could make the lab introduction more effective.

> *I found the presentation skills session most useful. It is very difficult to get used to standing up and talking in front of people. Receiving some training/advice was both helpful and comforting.*
>
> *I think it was a great course and it was very necessary . . . In this course, I learned how to present effectively and when it was appropriate to use PowerPoint, also how to interact with students who are struggling with concepts.*
>
> –Student feedback from the end-of-course evaluation survey

Alternative Scientific Careers

We have invited multiple guest speakers to give presentations on graduate and post-graduate professional opportunities. These guest speakers come from federal agencies, professional societies, and private biotechnology companies. The guest lectures demonstrate the breadth of career opportunities open to students with graduate degrees in the sciences. Following the presentations, we discuss how teaching and communication skills can provide graduate students with an advantage in obtaining and succeeding in such careers.

Interactive Game About Student Confidence

We sometimes conclude the course with an interactive activity in which GTAs assume the role of students and experience the impact of assessment on student confidence and engagement. This game also demonstrates the variety of potential student responses to a shared experience.

The course instructor draws a table on the board with a row for each class participant, a column for the participants' names, and ten columns corresponding to ten questions, as shown in Fig. 5.3. The instructor then offers instructions for the activity (see text box), and asks a student to repeat the instructions before the game begins. We do not share the topic of the questions prior to the activity, nor do we offer clues about the difficulty of the questions. The first five questions are very easy. Questions six, seven, and eight are more difficult, and nine is relatively easy. Question ten is almost impossible to solve. The questions, solutions, and explanations are shown in Table 5.3.

Fig. 5.3 Students completing the table in the interactive game

Table 5.3 Questions, solutions, and explanations for interactive game

	What is the next object in the pattern?	Solution	Explanation
1	1, 2, 3	4	Add 1
2	A, B, C	D	Next letter in alphabet
3	2, 7, 12, 17	22	Add 5
4	□ △ ◁▷ (triangle/square figures)	(four-triangle figure)	Attach a fourth triangle to the square
5	123, 234, 345	456	Add 1 to each digit
6	1/3, 4/1, −9	−16/3	The numerator is the square of the place in the pattern (e.g., 1^2, 2^2), the denominator is subtracted by 2
7	586, 6109, 73313	85658	Add 1 to the first digit, 2 to the second digit, 3 to the third digit...
8	1/2, 3/5, 5/8	7/11	Add 2 to each numerator, add 3 to each denominator
9	51, 62, 73	84	Add 1 to each digit
10	1933, 1934, 1959, 1962	1971	Birth years of course instructor's immediate family

Interactive Game Instructions

Participants will each write their name in one row on the table before beginning the activity, so that each row of the table corresponds to a class participant.

　　The activity requires students to complete ten questions. For each question, there are two envelopes: one that contains slips of paper with the first three or four numbers or characters in the pattern and a second envelope that contains slips of paper with the correct response (see Fig. 5.4). Prior to looking at each pattern, participants will estimate their likelihood of responding correctly (on a scale of 0 % to 100 %) and will write this under the appropriate question number in their row of the table. Participants then go to the envelope corresponding to the question, pull out a paper with the pattern, and attempt to solve it. Once they have a response, they then check their response against the correct answer in the corresponding response envelope. They then note whether their response was correct by writing a plus or a minus adjacent to their predicted likelihood of a correct response. Students then move to the second question, once again beginning with an estimate of their likelihood of figuring out the correct response. The process is repeated for all subsequent questions.

Fig. 5.4 Students looking at question and response envelopes while engaged in the interactive game

After all students have completed the game, we ask students to reflect on the game and its possible purposes. The follow-up discussion can include any or all of the following components, based on time available and the instructor's goals for the session.

Presentation of Instructions: At the beginning of the game, the presenter outlined the instructions and then asked a student to repeat them. We discuss alternative ways to reinforce and check students' understanding of instructions, such as writing instructions on the board, providing handouts, and questioning students.

Changes in Confidence Ratings: The differences in confidence ratings in the first column are indicative of differences amongst students. These differences may stem from student personalities as well as prior experiences and backgrounds, and these differences may influence how students approach the assignment. As the game progresses, students react differently to their successes and failures on individual questions. Some students change their confidence ratings based on how well they did on the prior questions, while others treat each question independently. We discuss how this situation applies in the classroom (e.g., when first-year students receive a lower than expected grade on their first test, or how the difficulty of a given item on an exam impacts student performance on subsequent items).

Student Investment in Class Activities: In the game, students differ in the amount of time and effort they invest in solving the questions, especially the more difficult questions. We discuss several measures that affect students' investment such as the weight of the grade in the course, personal work ethic, and insecurities.

An Unsolvable Question: The last question is virtually impossible for students to solve because it asks them to find the last number in an undecipherable pattern (years of birth of the instructor's immediate family). Some students struggle when faced with a question that they cannot easily solve, which may result in instances of cheating or giving up on the whole activity. In the earlier questions, there is no way to know if students are cheating. With this virtually unsolvable question, any correct answers are likely instances of cheating. In this question, we do not ask participants to report on the correctness of their response, but we raise the issue of cheating. We discuss what may lead students to cheat (e.g., an overachieving personality, insecurity, and pressure to get good grades to achieve career and graduate school goals).

Expert-Novice Knowledge Gap: Question ten also offers an opportunity to model the expert-novice gap. Students see the response to this question, but are unable to determine why this number would follow in the pattern. The instructor, on the other hand, finds the pattern to be obvious. She has background knowledge that the other participants do not have. We discuss how experts (graduate students, in this case) have much more background knowledge than novices (the undergraduates in their class). An awareness of the expert-novice knowledge gap can help instructors adapt their teaching to student background knowledge and position within the intended learning progression.

2-Credit Course for Graduate Students

This two-credit course provides students with a comprehensive understanding of theory and practice in science education. Our class meets once a week for an hour and forty minutes. We recommend that the class be held in a room with tables and movable chairs, which allows for group work and collaboration among the students. The course has two required textbooks: McKeachie and Svinicki (2014) *McKeachie's Teaching Tips* and Handlesman, Miller, and Pfund (2007) *Scientific Teaching* (hereafter referred to as *Teaching Tips* and *Scientific Teaching*, respectively). These are supplemented with papers from peer-reviewed journals and book chapters. We also offer students a list of recommended readings if they are interested in a deeper exploration of the topics.

Course sessions begin with a discussion of the homework, followed by presentations and activities related to key concepts in science education, as outlined below.

Session 1: Introduction to the Course

Icebreaker and Introduction

We open the first class session with an icebreaker activity. In this activity, each student gets a brown paper lunch bag and is tasked with introducing himself/herself using this bag. Students are also provided magazines, markers, scissors, and glue, and they generally use the bag and these materials to create a collage of sorts. They then introduce themselves using their brown bag creations, as shown in Figs. 5.5 and 5.6. This activity allows students to get to know one another beyond their scientific research interests. It also helps students to feel comfortable sharing information, which set the tone for the rest of the semester.

The course instructor then provides an overview of the syllabus and course expectations. Afterwards, all students complete a short survey so that the course can be better tailored to their needs and interests. This survey asks students why they took the course, what they expect to gain from it, and if they have suggestions for topics of interest.

Resources for Science Education

The instructor provides an overview of the many resources that are available in the area of science education. This overview introduces students to journals focusing on science education, conferences that generally include presentations on topics

Fig. 5.5 Students creating lunch bag collages in the icebreaker activity

Fig. 5.6 Students sharing their lunch bag collages in the icebreaker activity

in science education, and databases of education resources. See the TLC website (www.cmns-tlc.umd.edu/) for an up-to-date list of resources.

Homework

1. Read *Teaching Tips,* Chapter 3: Meeting a Class for the First Time. In one paragraph, describe your experience on your first day as a GTA.
2. Read *Scientific Teaching,* Chapter 1: Scientific Teaching. In one page, describe what resonated with you the most in this chapter and what you think you can use to improve your teaching.
3. Complete the Index of Learning Styles Questionnaire at www.engr.ncsu.edu/learningstyles/ilsweb.html (by Soloman & Felder). Print the results of the Index.

Session 2: Overview of Science Education

"Leading the Blind"

This activity is designed to demonstrate that, even when tasked with the same goal, instructors will use different approaches. Instructors may differ in their level of detail or whether they offer an overview or only focus on discrete steps or processes. Instructional style can also vary based on instructional goals. The activity also illustrates that different instructional styles vary in their effectiveness.

Leading the Blind Activity Instructions

Four students wait outside of the classroom. Three of these four students are given a blindfold to be used when they re-enter the room. The instructor then arranges a path or maze through the classroom that goes around, between, and even over desks and chairs, and shares it with all of the students inside the classroom.

One of the students from inside the class is designated as the leader and is instructed to get a blindfolded student from the beginning to the end point of the maze. All other students in the room take notes on the process that follows. The leader guides the blindfolded student through the maze in whichever way s/he chooses. When the blindfolded student reaches the end point, s/he removes the blindfold and is asked to re-walk the maze from the beginning to the end without the blindfold, but also without any guidance.

The student from outside who does not have a blindfold is brought into the class and designated as leader. S/he is shown the maze and asked to guide another blindfolded student through it. The maze activity is repeated, first with the blindfold and then without it.

In the third part of the activity, the student who just played the role of the blindfolded student is now designated as leader. This student already experienced the activity as a learner. As a result, s/he knows that s/he will not only be asked to guide the blindfolded student through the maze but that the learner will need to repeat the maze without guidance. The third blindfolded student is brought inside and the activity repeats.

The discussion of the activity generally highlights the following:

Leaders/instructors give different instructions: From our experience, in most cases the leader guides with verbal directions. We see a great deal of variation in the verbal instructions offered. Some leaders begin with a big-picture overview of the maze (e.g., "You will need to get across the room by going around chairs to get to the back of the classroom, come back to the front of the classroom, climb over a desk, and then zigzag between chairs to get to the far corner of the room."). Others focus on each unique component of the maze (e.g., "Go three steps forward."). In rare cases, the leader takes hold of the blindfolded student's hand and guides him/her through. These variations highlight differences in instructional style. The class discusses advantages and disadvantages of different instructional approaches in a real classroom situation.

Instructions change based on the known goals of the activity: In the third part of the activity, the leader knows that the blindfolded student will need to repeat the assignment and may adapt his/her instructions to this learning goal. Such instruction may include more of an articulation of the patterns that students will follow rather than instructions on each component of the maze. We discuss how teaching should reflect learning goals.

Experiences as a learner impact instructional approach: The leader in the third part of the activity previously experienced the activity as a learner. This prior experience may impact how s/he approaches the assignment, just as a classroom instructor's experience as a student may influence how s/he teaches.

Salient Topics in Science Education

The course instructor offers a brief overview of important issues and topics in science education. This overview introduces many of the topics that will be discussed in more detail in subsequent class meetings. This presentation generally covers an overview of why students leave the sciences and the leaky pipeline metaphor (Seymour & Hewitt, 1997), recommendations for reforming science education (AAMC-HHMI Committee, 2009; (AAAS, 2011; Handelsman et al., 2007), learning theories (e.g., constructivism), and an introduction to evidence-based teaching approaches (Handelsman et al., 2007).

> *Constructivism is probably the most significant theory we discussed. This really forces an instructor to create an atmosphere where students can build their own understanding (usually in an active learning activity) and not simply lecture and regurgitate.*
>
> –Student feedback from the end-of-course evaluation survey

"The Frustrated Professor" Case

This brief case study, taken from Handelsman et al. (2007, p. 127) presents a professor who is frustrated that students do not seem to be learning the material covered in lectures and the course textbook. Our class divides into small groups to discuss this case and respond to Handelsman et al.'s (2007) questions. This assignment serves as a summary of what was covered in the session and reinforces the importance of student-centered instructional approaches.

Homework

1. Read "Reaching the Second Tier: Learning and Teaching Styles in College Science Education" (Felder, 1993).
2. Read "Learning Styles and Strategies" (Felder & Soloman, www4.ncsu.edu/unity/lockers/users/f/felder/public/ILSdir/styles.htm)
3. Summarize in one page how you see yourself in term of Felder's learning style categories and the implications of this for your learning and teaching. Is this compatible with the profile that you received after taking the Index of Learning Styles questionnaire last week?

Session 3: Student Diversity in Learning Styles and Multiple Intelligences

Felder's Learning Styles

As Felder explains, people have different learning styles that impact how they best acquire new knowledge. He used the following questions to characterize student's learning styles (Felder, 1993, p. 287):

1. What type of information does the student preferentially perceive: *sensory*—sights, sounds, physical sensations, or *intuitive*—memories, ideas, insights?
2. Through which modality is sensory information most effectively perceived: *visual*—pictures, diagrams, graphs, demonstrations, or *verbal*—sounds, written and spoken words and formulas?
3. With which organization of information is the student most comfortable: *inductive*—facts and observations are given, underlying principles are inferred, or *deductive*—principles are given, consequences and applications are deduced?
4. How does the student prefer to process information: *actively*—through engagement in physical activity or discussion, or *reflectively*—through introspection?
5. How does the student progress toward understanding: *sequentially*—in a logical progression of small incremental steps, or *globally*—in large jumps, holistically?

We discuss Felder's learning styles and their implications for learning and teaching. Students may fall at one extreme or may have a more balanced learning style. Course instructors can adjust their teaching style to account for diversity in the classroom and respond to various learning preferences. Learners can also 'help themselves' by asking their instructors to provide information in a way that reflects their learning style or studying material in a way that capitalizes on how they learn best. See Felder (1993) and Felder and Silverman (1988) for specific examples.

Students share their results from the learning styles index and provide specific examples of how their learning style impacts both their learning and teaching. The course instructor provides examples of active learning approaches that cater to different learning styles.

Student Reflections on How their Learning Styles Impact their Learning and Teaching

Based on the questionnaire, I am well balanced between active and reflective styles of learning. As a graduate researcher, I enjoy the brainstorming sessions with my advisor and fellow lab mates a lot as we bounce off research ideas. But I also need time to reflect upon certain ideas and data before I draw conclusions from them.

If I had known the difference between sequential and global learners as an undergraduate, I would not have focused so much on solving problem after

(continued)

problem until I memorized how to come up with the answer. I would have instead focused more on the overall concepts and how to relate them to other subjects so that I would understand how to solve problems, rather than memorize.

I am very curious to know the distribution of learning styles in the course I am teaching. I will then be able to adjust my approach based on their responses, and this is both exciting and scary. I would expect to see many students with learning styles that favor the typical science lecture, and it would be a huge adjustment on my part if I found out many students in the class are creative and artistic ... Just being aware of these different styles and having the vocabulary to describe the differences makes clear to me how I can vary my approach. And, as is usually the case after reading about science education, I feel like I have so many things to improve!

–Excerpts from students' homework reflections

Gardner's Multiple Intelligences

To further illustrate student diversity in learning and abilities, the class engages in a discussion and an activity based on Gardner's theory of multiple intelligences (Gardner, 1983). Gardner posits that intelligence is not a single general ability but rather is a collection of different "modalities" of intelligence. Each person's intelligence is a composite of the multiple types of intelligence (task numbers refer to worksheet, described below):

Cognitive Intelligence

- Linguistic (task 8)
- Logical-Mathematical (tasks 1, 2, 7)

Artistic Intelligence

- Musical (task 3)
- Bodily-kinesthetic (task 4)
- Spatial (tasks 9)

Personal Intelligence

- Interpersonal (task 6)
- Intrapersonal (task 5)

People vary in their relative strength in each area. We discuss the implications for multiple intelligences in terms of classroom instruction. Gardner asserts that most academic courses value the cognitive intelligences above artistic and personal intelligences. Gardner recommends that teaching should not be restricted to specific types of intelligence, and that teaching to enhance all types of intelligence can empower learners.

Following this introduction to Gardner's theory of multiple intelligences, students divide into pairs and complete a series of short tasks. These tasks reflect Gardner's different types of intelligence.

Annotated Multiple Intelligences Worksheet

This worksheet was adapted from unpublished course materials developed by Tal Tamir, WIZO Haifa Academy of Design and Education, Israel.
Below we include solutions and discussion points for each task. The italicized text should not be included in the version of this worksheet that is distributed to students.

1. **The teacher asked the students to prove that $2 = 1$. Dan provided the following proof:**

   I started with the assumption that a = b (a valid assumption).
   I multiplied both sides of the equation with $a \Rightarrow a^2 = ab$
   I added $(a^2 - 2ab)$ to both sides $\Rightarrow a^2 + a^2 - 2ab = ab + a^2 - 2ab$
   $\Rightarrow 2a^2 - 2ab = a^2 - ab$
   $\Rightarrow 2(a^2 - ab) = a^2 - ab$
   I divided both sides with $(a^2 - ab) \Rightarrow 2 = 1$

   Please explain the $2 = 1$ paradox.
   Solution/Discussion: If $a = b$, then $a^2 - ab$ will equal zero. As a result, Dan divided by zero, which results in an undefined solution. This is one of many widely known mathematical paradoxes. For a description of this paradox, see Singh (1997, p. 294). Poundstone (1989) *also describes this and other paradoxes. This task challenges students' logical skills and mathematical knowledge.*

2. **Go through all the stars with only four lines without raising the pen from the paper.**

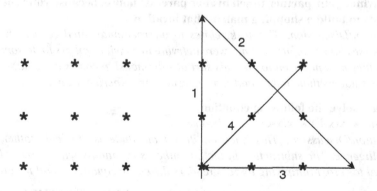

Solution/Discussion: This logic task challenges students to think "outside of the box." Students usually think only of potential solutions that respect the boundaries created by the dots, but it is impossible to complete the task with four lines that stay within these boundaries.

3. **Sing the lyrics of the first verse of the American national anthem using the melody of another song of your choice.**
   Solution/Discussion: This task tests students' musical intelligence. Some students may be able to complete the task easily, while others will struggle with the dissonance of attaching familiar lyrics to a different melody.

4. **Pat your stomach with your right hand in circle and, at the same time, tap your head.**
   Solution/Discussion: This task tests students' bodily-kinesthetic intelligence. As with the prior test, some students may be able to easily accomplish this task, while others may be unable to do so despite intense effort.

5. **Rank how well the following characterize you, using a 1–5 scale with (1) being the trait that is most characteristic of you, and (5) the trait that least characterizes you.**

   - **Honest**
   - **Loving**
   - **Hateful**
   - **Optimistic**
   - **Jealous**

   Solution/Discussion: This task requires intrapersonal intelligence. While there is no correct response, we discuss how students feel while they are completing the task. Some students have a strong awareness of how well these traits characterize them. Other students struggle to rank the five traits.

6. **Convince your partner to sell his/her parents' house, because your company wants to build a shopping mall at that location.**
   Solution/Discussion: This task relates to interpersonal intelligence. The class discusses how comfortable they were in trying to convince others to do something that they may not want to do. This sort of salesmanship comes more naturally to some students than others, and some may be more experienced in this sort of task.

7. **Please solve the following equation:**
   $x(x-a) * x(x-b) * x(x-c) \ldots \ldots x(x-z)=$
   Solution/Discussion: This question draws on students' logical-mathematical intelligence. The subtraction in the parentheses eventually will be x-x, which is equal to zero. Multiplying by zero renders the entire equation equal to zero.

8. **Find as many words as you can within the word GRANDFATHER.**
   Solution/Discussion: This task challenges students' linguistic intelligence. There are more than 500 words that can be formed from the letters in the word GRAND-FATHER. Students will vary in the number and complexity of words that they list

in response to this task, and they will also vary in how much they challenge themselves to find additional words. Some students are comfortable with a few words, while others will continue to search for all possible anagram solutions.

9. **Go to room 2249 in the Biology-Psychology Building.**
   Solution/Discussion: The final task addresses students' spatial intelligence. Room 2249 is in the building in which the class is taught; however, because of the complex layout of the building, it is generally difficult to find even for students who spend a lot of time in the building. We discuss how students find the room. Some look at the building map, others walk the halls in a trial-and-error approach, and some ask for directions either from other students in the class or other people in the building. Students also use logic and their knowledge of the way that rooms are numbered in the building (the thousands digit refers to the floor, the hundreds digit to the hall or wing, and the last two digits to the room's location within the wing).

Interpretation and Discussion

After students complete the worksheet, the whole class discusses this activity. This discussion generally covers the following topics:

Student diversity: Most students are more comfortable with some of the tasks than with others. A student may complete the logical-mathematical questions with ease, but have difficulty when asked to sing the national anthem to the tune of a different song. A different student may be less comfortable with the logical-mathematical tasks, but have no problem completing the musical task. We ask students to reflect on how they felt when completing each task.

Approach to the tasks: The course instructor asks each pair of students to describe their approach to the assignment. Some pairs work together on all tasks. Other students split the tasks, generally based on their comfort with and abilities in each area of intelligence, and work individually on their assigned portion. We discuss how these approaches play out in group work in the classroom and in labs, and whether course instructors should adjust to or promote different approaches to tasks.

If students are interested in learning more about Gardner's theory of multiple intelligences, we refer them to Smith (2008), which is available at www.infed.org/thinkers/gardner.htm and offers a concise overview of his theory.

Other Indices of Learning Styles
While we generally use activities based on Felder's learning styles and Gardner's multiple intelligences, other theories or indices could be used in this class session. Multiple indices of learning type or personality have

(continued)

been applied to student learning and classroom instruction. These include the Briggs-Meyers inventory (Lawrence, 2009), Fleming's VARK (www.vark-learn.com/english/index.asp), and Kolb's (1984) experiential learning styles. Regardless of the index used, these exercises are intended to raise awareness of diversity in how students prefer to learn and how instructional style impacts student learning. It also demonstrates how a theory can be both explained and illustrated through active learning approaches.

Homework

1. Prepare a presentation on an educational theorist: Working in small groups, students select an educational theorist, research that theorist (using one provided reference and selecting two additional references on their own), and prepare a ten-minute presentation for the class. This presentation should incorporate one or more active learning techniques.

Suggested Theorists and References

Jean Piaget: Huitt, W., & Hummel, J. (2003). Piaget's theory of cognitive development. *Educational psychology interactive*, *3*(2). Can be retrieved from www.edpsycinteractive.org/topics/cognition/piaget.html

Lev Vygotsky: www.funderstanding.com/content/vygotsky-and-social-cognition

John Dewey: www.infed.org/thinkers/et-dewey.htm

Jerome Bruner: www.infed.org/thinkers/bruner.htm

Benjamin Bloom: www.nwlink.com/~donclark/hrd/bloom.html

Lee Shulman: Shulman, L. S. (1986). Those Who Understand: Knowledge Growth in Teaching. Educational Researcher, 15(2), 4–31.

Richard E. Mayer: Mayer, R. E. (2002). Rote vs. Meaningful Learning, Theory into Practice, 41(4), 226–233.

Sessions 4 and 5: Theories in Science Education

Students give their presentations on important theorists in science education and explain how their theories have contributed to our current understanding of teaching and learning in science. Following each ten-minute presentation, other students in the class have the opportunity to ask questions. The instructor supplements each presentation with additional information if needed.

After each presentation, one student in the class is asked to repeat the key ideas from the presentation and the presenters verify that all key ideas were accurately recalled. The presenters then explain how they divided the work among themselves and the techniques they used to promote student recall of the key ideas in their presentation. All students in the class are then invited to provide feedback on the presentation and discuss the effectiveness of the techniques used. Students offer constructive criticism and discuss the pros and cons of different methods.

These presentations are loosely modeled on the jigsaw cooperative learning technique. The jigsaw technique was developed to foster cooperation in diverse classroom settings (Aronson, 1978). The technique makes students reliant on one another to learn a topic as they gain expertise in one component of the material and then share this expertise with other students, who have gained expertise in other components. The method typically involves dividing students first into specialized groups in which all students in the group gain expertise in an assigned topic (in this case, one theorist per group) and prepare to teach others about their area of expertise. Each expert group then splits up with each member going to a different table, so that the students are now grouped heterogeneously. In these heterogeneous groups, each member provides information on his/her area of expertise to complete the 'jigsaw puzzle' and fully cover the material. This method has been demonstrated to be effective in undergraduate science classes (Injaian, Smith, German Shipley, Marbach-Ad, & Fredericksen, 2011; Leman & Acar, 2012).

Splitting into multiple groups would be less effective given the small size that is typical of our class, so we adapted this approach to include smaller specialized groups and a single heterogeneous group that includes all class members. In a large class setting, our approach would be less feasible, and the classic jigsaw technique would be more efficient.

[Bruner's] spiral curriculum always stood out in my mind. Repetition is so important and this method allows you to introduce concepts in multiple layers of complexity.

I would definitely use [Bloom's taxonomy] to write assessments and present lectures in an organized way. It helps determine how deeply students understand a particular topic and if learning goals have been achieved. [It is a] very good organizational tool.

I really liked [Shulman's] theory that [we] presented on pedagogical content knowledge. Basically, this theory suggests that, in order to teach something, you need an extensive knowledge of both the content and the best way to teach that content. This class, for that reason, has been very helpful. I will use a lot of the tools I learned here in my future.

–Student feedback from the end-of-course evaluation survey

Homework

1. Read *Scientific Teaching*, Chapter 2: Active Learning and the article, "Can undergraduate biology students learn to ask higher level questions?" (Marbach-Ad & Sokolove, 2000a)
2. Choose five active learning approaches. In approximately one page, give an example for each of these approaches from your classroom experience (as a student or as an instructor).

Additional Resources on Active Learning

Eric Mazur's Confessions of a Converted Lecturer: www.youtube.com/watch?v=WwslBPj8GgI

Carl Wieman Science Education Initiative Clicker and Education Videos: www.cwsei.ubc.ca/resources/SEI_video.html

---

Session 6: Active Learning and Evidence-Based Teaching Approaches

Discussion of Approaches to Promote Active Learning

The course instructor provides an overview of different approaches to promote active learning, and students offer examples from their own experiences with these methods, either as students or as instructors. The presentation and discussion generally cover the following methods:

- Approaches to gauging student backgrounds

  - Brainstorming
  - Think-Pair-Share
  - Clicker questions
  - Problem-based learning
  - One minute essay/question
  - Concept mapping
  - Group discussion

- Approaches to delivering course content

  - Jigsaw
  - Decision making/debates
  - Problem-based learning/case studies
  - Concept mapping
  - Simulations
  - Role play
  - Field trips
  - Model-aided instruction

Research on Active Learning

The instructor introduces three studies assessing the effectiveness of specific active learning techniques in large-enrollment introductory biology classes:

1. Training students to ask higher-level questions (Marbach-Ad & Sokolove, 2000a)
2. Cooperative learning in studying for exams (Sokolove & Marbach-Ad, 1999)
3. Student-instructor interaction (Marbach-Ad & Sokolove, 2001)

These studies demonstrate that multiple techniques can be used to promote active learning and students benefit from active learning approaches at different stages in their learning process. These three cases are peer-reviewed research studies that compare classes that heavily employ active learning techniques to classes using traditional approaches. While these studies were chosen for their relevance to our population of students, other studies may be substituted.

[The course] has shifted my focus from teacher-centered lecture to a student-engaging atmosphere. We want to be good teachers, but that doesn't necessarily mean good lecturers .... [There are a] variety of active learning tools, so that you can pick/choose those that best fit your lecture layout or learning goals.

[A piece of useful information that I will take away from this course is] an understanding of what exactly active learning is: it's not just physically active. Also, [I learned] that there are many methods.

I am thinking of ways to incorporate more active learning exercises into my lab lectures next semester.

—Student feedback from the end-of-course evaluation survey

Categorizing Students' Questions

The class separates into small groups. Each group is given a short excerpt from an introductory science textbook and asked to write "their best questions" based on that material. After writing questions, the group then categorizes their own or another group's questions according to Marbach-Ad & Sokolove's taxonomy (2000a, 2000b).

Marbach-Ad and Sokolove (2000a, 2000b) developed a taxonomy to characterize student questions following a reading assignment. The taxonomy was created based on reading hundreds of questions that students wrote during the course of an introductory biology course. As a homework exercise, the course instructor asked the students, "... after reading this chapter, give me your best question" (p. 854). The instructor did not define what he meant by "best questions." The taxonomy derived from the analysis of these questions consists of seven categories (see box).

The questions and the taxonomy helped the instructors to (1) learn about students' prior knowledge and alternative conceptions, (2) identify topics that students were interested in learning, and (3) guide the students to write good questions.

There are many reasons to encourage students to write questions. First, the process of formulating questions is an important component in scientific research, and represents the inquiry nature of science. Second, student questions provide opportunity for the instructor to probe students' understanding and misunderstanding in a specific subject matter. Third, students' questions can increase the students' engagement with the material, which leads to better retention.

Following the creation of the taxonomy the instructor was able to define what he means by a "good question" and encouraged the students to write questions that reflect categories four through six, which involve more critical thinking. The taxonomy was used to provide evidence that students who were enrolled in an active learning version of the introductory course significantly improved the level of their questions over the semester compared to students who were enrolled in a more traditional, teacher-centered course taught by the same instructor.

Marbach-Ad and Sokolove's (2000b) Taxonomy

(0) Question is not logical or grammatical, is based on a basic misunderstanding or misconception, or does not fit in any other category

(1a) Question about a simple definition, concept, or fact that could be looked up in the textbook

(1b) Question about a more complex definition, concept, or fact explained fully in the textbook

(2) Ethical, moral, philosophical, or sociopolitical question; often begins with "why"

(3) Question for which the answer is a functional or evolutionary explanation; often begins with "why"

(4) Question in which the student seeks more information than is available in the textbook

(5) Question resulting from extended thought and synthesis of prior knowledge and information; often preceded by a summary, a paradox, or something puzzling

(6) Question that contains within it the kernel of a research hypothesis

Homework

1. Read *Scientific Teaching*: Chapter 3: Assessment. In no more than one page, reflect on the pros and cons of using rubrics.
2. Browse through the Field-tested Learning Assessment Guide (FLAG) website (www.flaguide.org/cat/cat.php). In no more than one page, describe two assessment tools that you have not previously used but would like to use in the future. Explain why.

Session 7: Assessment, Rubrics, and Grading

Overview of Assessment

Assessment is usually intended to provide feedback to students and instructors about the breadth and depth of student understanding (Handelsman et al., 2007). The instructor provides an overview of the use of assessment in science education, and highlights the importance of coordinating and integrating teaching, learning, and different modes of assessment (Barak & Dori, 2009). This overview generally includes the following:

Formative and summative assessment: Formative assessment provides ongoing feedback about student understanding. Summative assessment measures student progress and understanding at defined points.

Validity and reliability: While classroom assessments generally do not go through the same validation process as research instruments, instructors should still consider issues of validity and reliability in their assessments.

Item types: Assessments can include a variety of item types, such as open-ended or constructed response, multiple choice, fill-in-the-blank, and drawing or graphing.

Developing assessment items: When developing assessment items, instructors should consider the desired level of understanding and depth of the question according to Bloom's taxonomy (Bloom, 1984).

Writing Test and Quiz Items

Class participants are introduced to Bloom's taxonomy (Bloom, 1984). Students divide into small groups and are given a short excerpt from an introductory science textbook on a topic that is relevant to the classes they might teach. Each group creates at least three items to gauge student understanding of the material covered in the excerpt. The items should reflect different cognitive levels of Bloom's taxonomy (knowledge, understanding, application, analysis, synthesis, and evaluation). The class reconvenes to review and critique all of the items in terms of their quality, clarity, and cognitive level. Finally, the class revises a subset of these items.

Types of Assessment Tools

There are many different types of assessment tools. Classroom assessment may measure content knowledge, conceptual understanding, critical thinking, practical application, and/or attitudes and beliefs. Common assessment tools include exams with multiple choice and/or open-ended items, conceptual diagnostic tests, one-minute paper, clicker questions, performance assessments, and student assessment of learning gains (SALG). See the Field-tested Learning Assessment Guide (FLAG) website (www.flaguide.org/cat/cat.php) for more details about these and other assessment types. This activity on types of assessment tools can take one of two forms:

Option A—Instructor-led discussion: The instructor leads the class in a discussion of different types of assessment, using the FLAG website (www.flaguide.org/cat/cat.php) and/or the examples in *Scientific Teaching* (pp. 53–55) as a guide. The instructor draws on student homework in discussing the possible uses of each type of assessment tool.

Option B—Student presentations: An alternative way to introduce this topic is to have students lead the discussion on types of assessment. The instructor assigns each student one or two assessment tools to present, depending on the number of students in the course. Students prepare their presentations as homework prior to the class. In their presentations, students provide an overview of their assessment tool and show examples of how the assessment tool may be used in the context of the courses that they teach.

Grading

The instructor discusses the following topics related to grading:

Fairness and consistency in grading: The instructor offers tips on how to grade student work fairly and consistently. If time allows, the class can do the grading exercise from Session 3 of the mandatory 6-week GTA preparatory course (described on pages 137–138 in this book).

Graded, low-stakes, and ungraded assessments: Courses should include a variety of graded and ungraded assessments to provide feedback to both the students and the instructor. Using a combination of ungraded, low-stakes, and high-stakes assessments allows students to gauge their progress and allows instructors to collect ongoing feedback on student understanding without disproportionately impacting their final grades.

Item types: Assessments should include multiple item types. These different types allow students with different learning styles to demonstrate their understanding in different ways. The instructor discusses the advantages and disadvantages of different item types. For example, while multiple-choice items are easy to grade in terms of time required and ensuring objectivity, high quality multiple-choice items are difficult to develop and this tends to constrain the cognitive level of the question. Open-ended questions may be easier to write and more appropriate for measuring higher cognitive levels, but can be time-consuming to grade, particularly in large-enrollment classes. The grading of open-ended items can also be more subjective than the grading of multiple-choice items, and therefore the use of open-ended questions also should involve the concomitant development of a detailed rubric to help ensure consistency of grading.

Rubrics

The instructor introduces rubrics, provides examples, leads the class in a discussion of the pros and cons of using rubrics, and asks the students to identify classrooms contexts in which rubrics would be useful.

A rubric is a tool for communicating expectations about quality for specific tasks, activities, and assessments. It provides a set of criteria and standards that are typically linked to learning objectives. They may be used to outline learning goals, guide peer review, and grade assessments. Rubrics vary in their levels of specificity.

Rubrics are beneficial in that they (a) provide the student with a basis for self-evaluation and reflection on learning; (b) increase the student's understanding of the task and expectations about quality; and (c) facilitate accurate, objective, and fair assessment. However, rubrics may decrease the breadth or depth of the evaluation process by oversimplifying complex learning goals. Students may focus only on components or examples included in the rubric. In grading, instructors may only look for the components or examples included in the rubric and dismiss other relevant and correct responses that are not included in the rubric. Inappropriately designed rubrics can be more detrimental than beneficial to the learning process if they do not emphasize the desired or key elements that are being evaluated.

> *[The] most important thing I will take from this course is to use more feedback assessment. Students' attitudes surveys seem very powerful...I will make rubrics for assignments to give to students before the due date.*
>
> –Student feedback from the end-of-course evaluation survey

Mid-course Evaluation Survey

Session 7 is the midpoint of the course. This is an opportune time to conduct a mid-semester evaluation of student perception of the course to this point. In addition to providing feedback to the instructor, the mid-semester evaluation models the use of this type of assessment in the classroom. Our evaluation survey generally includes open-ended questions and Likert-style questions about specific components of the course.

Mid-semester Evaluation Survey

1. **Open-ended questions to gather student feedback**
   (a) Describe two types of useful information that you have taken away from this course.
   (b) Describe the two things that you liked the most about the course.
   (c) Describe two things that you would change in the course.

2. **Likert-style questions**

Please rate the effectiveness of the following course components.

	N/A	Was not effective	Slightly effective	Effective	Somewhat effective	Very effective
Reading assignments						
Homework assignments						
Group work outside of class						
In-class discussions						
In-class small group activities						
Introduction game (brown bags)						
Leading the blind activity						
Session on learning styles and multiple intelligences						
Presentations of learning theories						
Activity: categorizing students' questions						
Session on assessment						

Please elaborate on the course components that you rated as "not effective" or "slightly effective".

Homework

1. Read the articles "A misconception in biology: Amino acids and translation" (Fisher, 1983), "Cultural factors in the origin and remediation of alternative conceptions in physics" (Thijs & van den Berg, 1993), and "From misconception to concept inventory" (Marbach-Ad, 2009).
2. Write a two-page summary that addresses the following:

   (a) What is meant by 'misconception' (also known as alternative conceptions)
   (b) Five characteristics of misconceptions
   (c) Five possible reasons for students' misconceptions
   (d) One misconception that you had in the past
   (e) Two ways to help students to overcome misconceptions

Session 8: Students' Alternative Conceptions

This class session generally follows the format of the workshop entitled Overcoming Students' Alternative Conceptions in the Chemical and Life Sciences, which is detailed in Chap. 2.

Key Definitions

The instructor guides students in a brief discussion of the readings. This discussion highlights the definitions of the terms concept, conceptions, alternative conceptions, and preconceptions (Thijs & van den Berg, 1993). The class discusses characteristics of alternative conceptions. Alternative conceptions are widely shared, show similarities with views held in the historical development of science, and are persistent. Most alternative conceptions exist across many countries, within a variety of cultural and environmental contexts. However, other alternative conceptions may be dependent on culture.

Generative Causes of Alternative Conceptions

Worksheet: Students complete the following worksheet that includes questions that frequently elicit alternative conceptions.

Annotated Alternative Conceptions Worksheet

We give students this worksheet and ask them to respond individually to the questions before discussing their responses in small groups. These discussions raise awareness of some commonly held alternative conceptions. Below we include discussion points for each question. The italicized text should not be included in the version of this worksheet that is distributed to students.

Please complete this worksheet individually. After you have responded to all of the questions, in a small group discuss your responses, possible alternative conceptions for each question, and the generative causes that might lead to those alternative conceptions.

1. **Why do seasons happen? Why in the summer it is "hot" and in the winter it is "cold"?**

   Discussion points: A commonly held alternative conception is that seasons occur due to variations in the distance between the Earth and sun. This alternative conception may result in part from perspective drawings that illustrate the Earth's orbit as highly elliptical. Another possible origin of the alternative conception is experience in everyday life with small heat sources such as fireplaces or candles. Students have experienced that they feel the heat from sources most strongly when they are close to the source, and that the heat dissipates as they go farther from the source.

   The video "A Private Universe: Minds of our Own" (www.learner.org/ resources/series28.html), produced by the Harvard-Smithsonian Center for Astrophysics, demonstrates the ubiquity of this alternative conception and offers explanations for some of the sources of the alternative conception.

2. **Why do we have different phases of the moon?**

   Discussion points: A commonly held alternative conception is that phases of the moon result from a shadow of the Earth or clouds, or from the Earth's or the moon's rotation. These alternative conceptions may stem from overgeneraliza- tion of the eclipse phenomena and/or experience with shadows in everyday life.

   The video "A Private Universe: Minds of our Own" also discusses students' false explanation for the different phases of the moon. This and other alternative conceptions related to the moon can be found at moon.nasa.gov/ moonmisconceptions.cfm.

3. **Why, in hospitals, might nurses take the plants from patient rooms at night?**

   Discussion points: Some nurses may take plants from patients' rooms at night because they fear the plants could deplete the oxygen in the room while adding CO_2. During the day, plants emit more oxygen through photosynthesis than they deplete through respiration. At night, photosynthesis is limited, so the net effect on oxygen is negative and the effect on CO_2 is positive at night—but this impact is negligible, and more than compensated by the daytime impact.

   This question raises several common alternative conceptions that students may hold. First, many students confuse respiration with photosynthesis, and think that only animals engage in respiration and plants only engage in photosynthesis. A second alternative conception is the belief that photosynthesis only occurs during the day and respiration only occurs during the night. Third, students may hold alternative conceptions about plants' overall impact on air composition as a result of the combination of the photosynthesis and respiration processes. Over- simplification of how scientific processes play out can lead to these alternative conceptions.

Younger students may also hold an alternative conception that plants are not living things because they do not move or behave as animals. This alternative conception is held to varying degrees in different countries, due to cultural and linguistic differences (Stavy & Wax, 1989).

4. **The balloon was left out in the sun. Why did the balloon pop?**

Discussion points: A commonly held alternative conception, especially among young learners, is that a balloon will pop in the sun due to the impact of heat on the balloon material. This alternative conception may stem from a lack of knowledge about particle theory and how air molecules behave at different temperatures. Some alternative conceptions derived initially from a lack of knowledge may persist even as knowledge is gained, because students continue to rely on their intuition about what they see rather than applying their new knowledge.

5. **Consider a copper wire. Divide it into two equal parts. Divide one half into two equal parts. Continue dividing in the same way. Will this process come to an end?**

Discussion points: Some students hold the alternative conception that the process will not come to an end. This alternative conception results from an over-generalization of the mathematical concept of infinity. Research indicates that students intuitively answer this question correctly before they are formally taught the concept of infinity. In early high school, around the time that most students learn about infinity, the prevalence of this alternative conception peaks as students erroneously apply this mathematical concept to the material world (Stavy & Tirosh, 2000; Tirosh, Stavy, & Aboulafia, 1998).

6. A. **Two roommates fall ill: one has an ear infection and one has pneumonia. Is it possible that the same causative agent is responsible for both types of disease?**

   (a) **Yes, because both individuals live in the same room and therefore the source of the infection has to be the same.**
   (b) **Yes, because the same bacteria can adapt to different surroundings.**
   (c) **No, because each bacterium would cause one specific disease.**
   (d) **No, because one infection is in the lung while the other is in the ear.**
   (e) **I do not know the answer to this question.**

   B. **Explain your response.**

   Discussion points: This question was developed as a part of the Host-Pathogen Interaction (HPI) concept inventory to assess students' understanding of various HPI topics in a sequence of undergraduate courses. An analysis of student responses indicated that students in lower-level courses generally choose the correct answer (b) based on their intuition and experiences in everyday life. However, students who took an upper-level course on microbial pathogenesis were less likely to provide the correct response. Students in this course had learned about some bacteria that are specific to one type of disease, and were more likely to over-generalize this knowledge and choose response d. For more details about the concept inventory and findings from the analysis of student responses, see Marbach-Ad, et al. (2010).

7. **Two sugar cubes are added to a bowl containing 20 ounces of water. One sugar cube is added to a bowl containing 10 ounces of water. Which bowl of water is sweeter?**

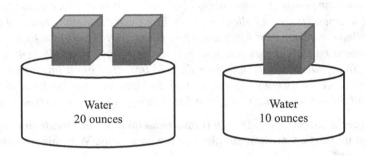

Discussion points: While most university students see this question and know that the bowls of water are equally sweet, younger students may say that the bowl with two sugar cubes is sweeter (Stavy, Strauss, Orpaz, & Carmi, 1982). *This response can be the result of confusing extensive and intensive properties. Extensive and intensive properties are classifications for physical properties that differ from one another in that extensive properties depend on the system size or amount of material in the system, while intensive properties do not. For this question, students need to understand that concentration is an extensive property in that it depends on the proportion of materials (water, sugar) in the system. Young students may rely on intuition and neglect to evaluate all variables in the problem and focus on one variable, such as the amount of sugar. At the university level, students tend to hold alternative conceptions about extensive and intensive properties in more complex problems, such as the difference between temperature (intensive) and heat (extensive), or weight (intensive) and mass (extensive).*

8. **Cats usually have five digits on each paw. Omer has a cat with six digits. Omer's cat gave birth to two kittens that also had six digits. How can you explain this? Was something transmitted from the mother to the kittens? If so, what is it?** (Marbach-Ad & Stavy, 2000)

   Discussion points: This question requires students to make connections between the macro-level (the number of digits trait) and the micro-level (the genetic material passed from the mother to the kittens). This question was used in a research study, which showed that high school and university students had difficulty explaining genetic phenomena on the macro-level by using cellular and molecular (micro-level) explanations (Marbach-Ad & Stavy, 2000).

   University students who are asked to make connections between macro- and micro-levels often have to do so for content that involves making interdisciplinary connections, which introduces additional complications. For example, micro-level interpretations of biological phenomena often involve explanations from both the physics and chemistry domains (e.g., molecular configurations).

While graduate students in the sciences are unlikely to hold alternative conceptions for some of the questions, the worksheet serves as the foundation for discussion about alternative conceptions, their characteristics, and their generative causes. Students first complete the worksheet individually and then discuss their responses and possible alternative conceptions in small groups.

Video: After these small group discussions, the class watches a scene from the video, "A Private Universe: Minds of our Own." This scene summarizes characteristics and possible generative causes for students' alternative conceptions in astronomy, as documented through interviews with Harvard graduates, their professors, and a bright ninth-grader who has some confused ideas about the orbits of the planets. The video can be accessed at www.learner.org/resources/series28.html?pop=yes& pid=9#.

Discussion: The instructor leads the class in a discussion about misconceptions and potential generative causes of the alternative conceptions (e.g., overgeneralization, oversimplification, and use of misleading intuitions or experiences from everyday life). Table 5.4 provides a list of generative causes of alternative conceptions and the related questions from the worksheet.

Identifying and Overcoming Alternative Conceptions

The class discusses various methods that can be used by classroom instructors to identify and overcome students' alternative conceptions, including those that were suggested by the graduate students in their homework assignments. Commonly used ways to identify alternative conceptions include having students answer open-ended questions (e.g., fill-in-the blank, essays), paraphrase concepts, and create graphical organizers (e.g., concept maps, word associations, charts, drawings).

When alternative conceptions have been identified, teachers can address them in different ways. Telling students that they do not understand the concept is usually

Table 5.4 Generative causes of alternative conceptions and related questions from alternative conceptions worksheet

Generative causes of alternative conceptions	Question(s)
Everyday language/experience	**1, 2, 3**
Intuition	**4, 7**
Over-generalization	**2, 5, 6**
Under-generalization or over-simplification	**3**
Confusion between extensive and intensive entities	**7**
Connections between the macro- and micro-level	**8**

insufficient. Methods that are more effective in overcoming students alternative conceptions include asking students to explain processes, using graphical organizers (e.g., models and concepts maps), using examples from the history of science to illustrate how our understanding of concepts developed over time, and creating situations that force students to try to resolve the conflict between their alternative conception and the scientific concept.

> *In reading the article by Thijs and Van Den Berg (1993), I found that even though I've had both a year of high school and year of college level physics classes, I still hold many common physics misconceptions.*
>
> *I encourage students to discuss the topic in my presence so that I can identify individuals who are having trouble in understanding the concept. Sometimes I will encourage their friends (who have understood the topic) to explain it to students who are having trouble ... Moreover, I make it a point to carry 3D models or other supporting material to class, which might help students to visualize the concept better in their heads.*
>
> —Excerpts from students' homework reflections

Homework

1. Find and read a paper about the impact of using concept maps or models on student achievement. Write a one-page summary responding to the following questions.

   - What did you find most interesting in the paper?
   - What assessment tools were used in the study, and what are the pros and cons of each tool?
   - What would you change in terms of the research methodology used to evaluate the impact of the model or concept map?

   Come to class prepared to describe your paper in a few sentences.

Session 9: Using Visual Representations—Concept Maps and Models

Overview of Visual Representations

The instructor provides a brief overview of visual representations and their role in science education. This overview focuses on concept maps and models, which are visual representations that have been demonstrated to be useful in improving undergraduate science education (Marbach-Ad, Rotbain, & Stavy, 2005; Marbach-Ad, Yarden, & Gershoni, 2007; Yarden, Marbach-Ad, & Gershoni, 2004).

Concept maps "represent meaningful relationships between concepts in the form of propositions. *Propositions* are two or more concept labels linked by words in a semantic unit" (Novak & Gowin, 1984 p. 15). Concept maps provide a way of capturing students' understanding of a key concept and related concepts.

Models are representations of "ideas, objects, events, systems or processes" (Gilbert, Boulter, & Rutherford, 1998, p. 92). They encompass physical objects, as well as verbal and mathematical descriptions. In science education, didactic or teaching models are used to aid students in developing an understanding of scientific phenomena and concepts through a cognitively appropriate representation. (Rotbain, Marbach-Ad, & Stavy, 2006).

Examples of the Effective Use of Models and Concept Maps

Students share the papers they selected for their homework assignment. For each paper, they discuss and critique how a model or concept map was used, the effectiveness of this intervention, and the research methods used to evaluate effectiveness. Students then discuss how they have used or could use concept maps, models, or other visual representations in their own teaching.

I selected this paper because, when I teach general chemistry next year, I know that at least half of my students will be engineers and the vast majority of them do not want to take chemistry ... What I loved about this study was that they engaged the students in learning by showing them that the subjects are interrelated. They took concepts and showed the connection between chemistry and engineering and physics and beyond, which in turn allowed the students to understand that there is a relationship between all their subjects and that they all build on top of each other.

The instruction in the two experimental and control groups in this study was done by different teachers, which I would change so that both groups of students would receive instruction from the same person.

I would like to see more research on the retention of the knowledge that the students gained from the pilot course. The researchers do show a higher average GPA for those who took the pilot course versus those that did not after their first and second years ... [but this] does not actually mean they retained anything they learned from the course.

–Excerpts from students' homework reflections

This discussion provides an opportunity to discuss not only concept maps and models, but also science education literature and how this body of literature can serve as a valuable resource for instructors. Students were impressed by the accessibility of an extensive body of literature on specific instructional techniques, and the ease with which they could adapt these techniques to their own teaching.

A Comparative Analysis of Different Types of Models

The instructor shares a study assessing the effectiveness of three types of models in teaching genetics: a three dimensional model (bead model), computer simulation, and illustrations like those frequently found in genetics textbooks. The experimental design included similar activities for each type of model, with three different groups of students experiencing each model/activity and a fourth group of students serving as a control group that experiencing traditional lecture. The three instructional approaches that incorporated models were all more effective than traditional lecturing. The effectiveness of different types of models depended on the content (e.g., computer simulations were best for teaching dynamic processes; bead models were best for representing structure). Furthermore, the research findings suggested that open-ended assessment items can illuminate differences in student understanding that may not be captured by multiple choice assessment items. See Rotbein, Stavy, & Marbach-Ad (2008) for a summary of the study; more detailed descriptions of the interventions and research findings can be found in Rotbein, Marbach-Ad, & Stavy (2005; Rotbain et al., 2006, Rotbein, Marbach-Ad, & Stavy, 2008) and Marbach-Ad, Rotbain, & Stavy (2005, 2008).

Developing and Evaluating Concept Maps

The class divides into small groups of two or three students to develop their own concept map that will be evaluated according to the scoring criteria outlined in Novak and Gowin (1984, pp. 36–37). This activity exposes students to both the process of creating a concept map and a methodology for evaluating concept maps.

Each group is given the same set of 11 concepts (living things, plants, animals, molecules, water, motion, states, liquid, solid, gas, and heat), with each concept written on a separate card or sticky note. The groups review Novak and Gowin's (1984) methodology and criteria for scoring concept maps. These criteria relate to the validity of the relationships between the terms, the hierarchy of the concepts, the cross-links between the segments of the concept hierarchy, and the appropriateness of the specific events or objects that are used as examples. Each group seeks to develop a concept map that achieves a high score according to these criteria. To create the concept maps, students arrange the cards or sticky notes on a large piece of paper, placing them in a logical order that allows them to create as many connections as possible between concepts. They then draw lines between related pairs of concepts and add propositions that describe the relationships between the two concepts. When they are done, they exchange concept maps with another group and use Novak and Gowin's criteria to score the other group's concept map.

The class reconvenes to discuss the pros and cons of having students create concept maps and using Novak and Gowin's criteria for evaluating them. Class participants suggest different ways that they could implement concept maps in their courses and propose alternative approaches for evaluating students' concept maps. They also weigh the benefits of using concept mapping against the time necessary for undergraduates to learn how to create concept maps, especially the time required to develop the detailed, multi-leveled concept maps that would achieve a high score using Novak and Gowin's criteria.

I appreciated concept mapping because it can include other learning theories such as Bloom's taxonomy. In future teaching, concept mapping will serve as a self-assessment tool for my students, and allow me to observe their progress in the topic.

The theory that emphasized the importance of visual learning will certainly be applied in my future teaching as I will try to use models, drawings, and pictures as much as possible.

–Student feedback from the end-of-course evaluation survey

Homework

1. Read *Teaching Tips,* Chapter 17: Technology and Teaching
2. A guest speaker from the university's Division of Information Technology will give a presentation on implementing new technologies in science classes. Prepare three questions to ask the presenter.
3. *Optional assignment*: View a video on using technology to enhance undergraduate teaching (e.g., TED Talks). Given the rapid pace of technological innovation, we suggest looking for recent videos that capture up-to-date educational technologies.

Session 10: Using Technology in Science Education

The landscape of technology uses in science education is rapidly changing. For this reason, we generally invite a guest who is an expert in the latest innovations in educational technology to lead this session. The guest speaker brings examples from courses throughout the university and focuses on examples from our science departments. S/he is able to discuss the spectrum of technology use in science classrooms at the university, with some classes using little technology and others employing multiple innovative technologies in flipped or blended learning environments. To model technology-aided instruction, this class session uses clickers and students are asked to provide responses throughout the presentation by voting with their clickers. The presentation typically takes a multimedia format with embedded videos, audio clips, and/or other media to demonstrate the range of possibilities.

An Overview of the Use of Technology in Science Education

The guest speaker introduces different types of technology-aided instruction that are commonly used in science education, and offers students an opportunity to discuss the pros and cons of using each of these instructional technologies. The following are suggested activities to engage students in these kinds of conversations:

Controversies regarding using technologies in the classroom: Students are polled via clicker about different uses of technology in the classroom that may be controversial, such as using laptops or cell phones. A visualization of poll responses is shared on the projector. The class then discusses ways to capture the benefits of these technologies while mediating the impact of their potential misuse.

Good and bad experiences with technology in the classroom: The class divides into groups of two or three students. Each student provides one example of a good use of technology and one example of a bad use of technology from their experience in the classroom. The group then reconvenes and each group shares selected examples.

Student diversity and technology: The class discusses how different student characteristics, such as traditional and non-traditional status, commuter status, year of study, and their technological competence, can affect the use of instructional technology.

Potential Uses of Technology in Science Education

Promoting engagement: Use technologies such as Facebook, Twitter, YouTube, blogs, wikis, and clickers to promote interaction among students and with the instructor both during and outside of class.

Facilitating group work: Use multiuser software such as wikis, blogs, Google apps, and tools within the university's learning management system to create and share content. Students can collaborate in creating content, engage in peer review of others' content, and discuss course content with classmates.

Using online content: Use online content, such as animations and presentations, to supplement or even replace traditional, in-class lectures. Instructors may create original resources or use those are freely available from a variety of outside organizations, including TED-Ed (www.ed.ted.com), the KhanAcademy (www.khanacademy.org), and Coursera (www.coursera.org).

(continued)

Creating student presentations: Students can create slide shows, posters, or multimedia presentations through Google apps, presentation software (Prezi, PowerPoint, Keynote), blogs, YouTube, or workspaces on the course management system.

Case Study—Use of Technology in the Classroom

We invite a faculty member who has successfully implemented innovative technology in his/her science classroom to lead this part of the session. To make this presentation particularly relevant, we try to invite a faculty member who teaches in the same discipline as most or all of the graduate students in the class. This allows the presenter to discuss technology and teaching from a PCK perspective. The class concludes with a question-and-answer session.

Session 11: Observation of an Undergraduate Science Class

Observing an effective instructor allows the class to see in practice many of the instructional techniques they have learned about over the course of the semester. In this session, we generally bring the whole class to observe an undergraduate science class. We have been fortunate that our class meets at the same time an introductory biology class taught by an instructor who employs many evidence-based teaching approaches. With the prior approval of this instructor, our students sit in on the introductory biology class. During the class, they complete an observation rubric and take informal notes on the teaching. If it is not feasible to bring an entire class to conduct an observation, students may observe a class on their own.

Observation Rubric

There are many types of observation rubrics. Because students conduct the observation during the first half of the class period and must be prepared to reflect on the observation immediately after it occurs, we provide them with a simplified observation rubric. If students complete the observation independently and outside of class time, a more comprehensive rubric may be appropriate. Students may also benefit from using the same observation rubric as is used for the departmental peer review process, if one exists.

Discussion of Observations

After observing the undergraduate class, the group reconvenes to reflect on the observation. The observation rubric can guide this discussion as students share their responses to the items on the rubric. Students are also encouraged to share what they learned from the observation, highlight the strengths and weaknesses of the teaching, and suggest alternate instructional approaches that might have been effective.

Homework—Final Project

Students will develop a "teachable unit" as their final project. The teachable unit should include aligned learning goals, activities, and assessment. This assignment can be done individually or in small groups, depending on the size of the class and the amount of time available for the culminating project presentations. In our course, students present their teachable units to the class in Sessions 12–14. The expectation is that the teachable unit will take students more than one week to develop, so it should be introduced to the class and discussed several weeks before the due date so that students have ample time to prepare.

The class instructor invites students to submit a draft version of their teachable unit for feedback at any point up to three days prior to the Session 12 class meeting. As an additional resource, students are given an exemplary teachable unit created by a graduate student in a previous class.

1. Read *Scientific Teaching,* Chapter 5: A Framework for Constructing a Teachable Unit; and the articles "Money, sex, and drugs: A case study to teach the genetics of antibiotic resistance" (Cloud-Hansen, Kuehner, Tong, Miller, & Handelsman, 2008) and "Aligning goals, assessments, and activities: An approach to teaching PCR and gel electrophoresis" (Phillips, Robertson, Batzli, Harris, & Miller, 2008).
2. Plan a teachable unit to present to the class during one of the last three class meetings.

Instructions for Teachable Unit

1. Designate the topic and the parameters for teaching it (e.g., course, topic, timeframe). You may choose a topic that you have taught previously or a topic that you plan to teach in the future.
2. Build a table that includes learning goals, activities, assessments, and alignment (similar to Table 5.3 p. 89 in *Scientific Teaching*).

(continued)

3. Create a teaching plan that describes how the content will be taught and assessed. You may want to include timeframes for components of the unit, supplies needed, instructions for the students and/or instructions for instructors (e.g., instructor's notes).
4. Include supplemental materials, such as an explanation of activities, handouts/worksheet for students, and assessment tools for measuring the achievement of learning goals.
5. Explain how the unit addresses the potential diversity of students in the class.
6. Prepare a presentation of no more than ten minutes to share your teachable unit. Before class, upload your completed assignment to the course learning management system. If you have materials you want to distribute to the class, make a few copies to bring to class or email the materials to everyone before class.

Sessions 12–14: Presentations of Teachable Units

Students give ten-minute presentations on their teachable units. If possible, the presentation should incorporate one of the activities from the teachable unit. During the presentation, students in the audience complete a Review Rubric for Teachable Units (*Scientific Teaching,* pp. 86–87) for each presentation. At the end of the presentation, these scored rubrics are given to the presenter. Following each presentation, the other students in the class have five minutes to reflect on the teachable unit and the presentation.

While students are instructed to convey information clearly and succinctly, they are encouraged to incorporate a short activity in their presentation (e.g., handing out a worksheet, directing a game or role-play, or engaging students in an experiment). Students may use presentation software if they wish, but they are not required to do so. Depending on the number of presentations, this activity generally takes most of Sessions 12, 13, and 14.

The thing I will take away the most from the course is the teachable unit because it took all the concepts from the course and put it into a useable format. I also took away the diversity of teaching approaches that are available and how each of them has their own strengths and weaknesses.

I will construct all future lecture notes as teachable units.

(continued)

> *I enjoyed creating the teachable unit. I chose a topic I care very much about, and this assignment gave me the opportunity to develop a plan to address it.*
>
> *I used [PCK] while constructing my teachable unit. This theory just made sense to me. I have had similar experiences where I knew a topic very well, but for teaching that topic for someone else, I had to rethink everything I knew. So this way, the theory will be especially useful when I make more teachable units.*
>
> –Student feedback from the end-of-course evaluation survey

End-of-Course Evaluation Survey

In the last class session, time should be reserved to solicit feedback from the students on the class and to have them complete an evaluation survey. This survey is independent of the university course evaluation system and is intended to provide targeted feedback to the course instructor. Because the course curriculum is flexible, future offerings of the courses can be adjusted in response to student feedback. Our end-of-course evaluation survey generally includes the following questions, with additional questions added as necessary to gather feedback on specific course components of interest.

1. Why did you choose to take this course?
2. Please select one learning theory that we covered in the course and, in a few sentences, explain how it may be applicable in your current or future teaching.
3. Describe two pieces of useful information that you will take away from this course.
4. Describe two things that you liked the most about the course.
5. Describe two things that you would change in the course.
6. Has this course impacted how you teach? If so, how?
7. Do you think that all science graduate students should be required to take this course? Why or why not?
8. Would you recommend this course to all your peers? A particular subset of your peers (e.g., those who plan to concentrate on teaching-related careers)? Why or why not?
9. Has this course impacted your career interests? If so, how?

The purpose of the end-of-semester survey is to receive feedback from students, and see if we achieved our overarching goals for the course. For example, since the students are not science education majors, we are not expecting that they will remember all the theories that we discussed in the course; however, we want them be able to discuss at least one theory and explain how they will utilize this theory in the classroom. It could also be very informative to ask students again about specific activities, similar to the way that we asked in the mid-evaluation survey (pp. 163–164). Our survey also includes a question asking the students for their informed consent to analyze their responses for research purposes. Their consent allows us to use their responses in published research as well as to inform future course offerings.

References

AAMC-HHMI Committee. (2009). *Scientific foundations for future physicians*. Washington, DC: Association of American Medical Colleges. members.aamc.org/eweb/upload/Scientific%20Foundations%20for%20Future%20Physicians%20%20Report2%202009.pdf

Addy, T. M., & Blanchard, M. R. (2010). The problem with reform from the bottom up: Instructional practices and teacher beliefs of graduate teaching assistants following a reform-minded university teacher certificate programme. *International Journal of Science Education, 32*(8), 1045–1071.

Alvine, A., Judson, T. W., Schein, M., & Yoshida, T. (2007). What graduate students (and the rest of us) can learn from lesson study. *College Teaching, 55*(3), 109–113.

American Association for the Advancement of Science (AAAS). (2011). *Vision and change: A call to action*. Washington, DC: AAAS.

Aronson, E. (1978). *The jigsaw classroom*. Oxford, England: Sage.

Austin, A. E. (2002). Preparing the next generation of faculty: Graduate school as socialization to the academic career. *The Journal of Higher Education, 73*(1), 94–122.

Austin, A. E. (2011). *Promoting evidence-based change in undergraduate science education*. A paper commissioned by the National Academies National Research Council Board on Science Education. dev.tidemarkinstitute.org/sites/default/files/documents/Use%20of%20Evidence%20in%20Changinge%20Undergraduate%20Science%20Education%20%28Austin%29.pdf

Austin, A. E., Campa, H., Pfund, C., Gillian-Daniel, D. L., Mathieu, R., & Stoddart, J. (2009). Preparing STEM doctoral students for future faculty careers. *New Directions for Teaching and Learning, 117,* 83–95.

Austin, A. E., Connolly, M. R., & Colbeck, C. L. (2008). Strategies for preparing integrated faculty: The center for the integration of research, teaching, and learning. *New Directions for Teaching and Learning, 113,* 69–81.

Austin, A. E., & McDaniels, M. (2006). Preparing the professoriate of the future: Graduate student socialization for faculty roles. In J. C. Smart (Ed.), *Handbook of theory and research*. New York, NY: Springer.

Barak, M., & Dori, Y. J. (2009). Enhancing higher order thinking skills among inservice teachers via embedded assessment. *Journal of Science Teacher Education, 20,* 459–474.

Bloom, B. S. (1984). *Taxonomy of educational objectives: Handbook 1: Cognitive domain*. New York, NY: Longman Inc.

Bond-Robinson, J., & Rodriques, R. (2006). Catalyzing graduate teaching assistants' laboratory teaching through research. *Journal of Chemical Education, 83*(2), 313–323.

Bouwma-Gearhart, J., Millar, S., Barger, S., & Connolly, M. (2007). *Doctoral and postdoctoral STEM teaching-related professional development: Effects on training and early career periods*. Paper presented at the Annual Meeting of the American Educational Research Association, Chicago, IL. Retrieved from www.cirtl.net/files/Bouwma-Gearhart_Doctoral%20and%20Postdoctoral%20STEM_2007.pdf

Boyer Commission on Educating Undergraduates in the Research University. (1998). *Reinventing undergraduate education: A blueprint for America's research universities*. Stony Brook, NY: State University of New York at Stony Brook.

Bretz, L. S. (2005). All students are not created equal: Learning styles in the chemistry classroom. In N. Pienta, M. Cooper, & T. Greenbowe (Eds.), *Chemist's guide to effective teaching* (pp.28–40). Upper Saddle River, NJ: Pearson Prentice Hall.

Brownell, S. E., & Tanner, K. D. (2012). Barriers to faculty pedagogical change: Lack of training, time, incentives, and . . . tensions with professional identity? *CBE Life Sciences Education, 11,* 339–346.

Caserio, M., Coppola, B. P., Lichter, R. L., Bentley, A. K., Bowman, M. D., Mangham, A. N., . . . Seeman, J. I. (2004). Responses to changing needs in U.S. doctoral education. *Journal of Chemical Education, 81,* 1698–1703.

Cloud-Hansen, K. A., Kuehner, J. N., Tong, L. L., Miller, S., & Handelsman, J. (2008). Money, sex, and drugs: A case study to teach the genetics of antibiotic resistance. *Cbe-Life Sciences Education, 7*(3), 302–309. doi:10.1187/cbe.07-12-0099

DeChenne, S. E., Enochs, L. G., & Needham, M. (2012). Science, technology, engineering, and mathematics graduate teaching assistants teaching self-Efficacy. *Journal of the Scholarship of Teaching and Learning, 12*(4), 102–123.

Dotger, S. (2011). Exploring and developing graduate teaching assistants' pedagogies via lesson study. *Teaching in Higher Education, 16*(2), 157–169.

Felder, R. M. (1993). Learning and teaching styles in college science education. *Journal of College Science Teaching, 23*(5), 286–290.

Felder, R. M., & Silverman, L. K. (1988). Learning and teaching styles in engineering education. *Engineering Education, 78*(7), 674–681.

Feldon, D. F., Peugh, J., Timmerman, B. E., Maher, M. A., Hurst, M., Strickland, D., . . . Stiegelmeyer, C. (2011). Graduate students' teaching experiences improve their methodological research skills. *Science, 333*(6045), 1037–1039. doi:10.1126/science.1204109.

Fisher, K. M. (1983). *Amino acids and translation: a misconception in biology.* Paper presented at the International Seminar on Misconceptions in Science and Mathematics, Cornell University, Ithaca, NY.

French, D., & Russell, C. (2002). Do graduate teaching assistants benefit from teaching inquiry-based laboratories? *Bioscience, 52*(11), 1036–1041.

Gardner, G. E., & Jones, M. G. (2011). Pedagogical preparation of the science graduate teaching assistant: Challenges and Implications. *Science Educator, 20*(2), 31–41.

Gardner, H. (1983). *Frames of mind: The theory of multiple intelligences.* New York, NY: Basic books.

Gilbert, J. K., Boulter, C., & Rutherford, M. (1998). Models in explanations, Part 1: Horses for courses? *International Journal of Science Education, 20*(1), 83–97.

Golde, C. M., & Dore, T. M. (2001). *At cross purposes: What the experiences of doctoral students reveal about doctoral education.* Philadelphia, PA: Pew Charitable Trusts. Retrieved from www.phd-survey.org

Greenler, R., & Barnicle, K. (2011). *Building an online professional learning community.* Paper presented at the 27th Annual Conference on Distance Teaching & Learning, Madison, WI. Retrieved from www.uwex.edu/disted/conference/Resource_library/proceedings/46079_2011. pdf

Handelsman, J., Ebert-May, D., Beichner, R., Bruns, P., Chang, A., DeHaan, R., . . . Wood, W. B. (2004). Scientific teaching. *Science, 304*(5670), 521–522.

Handelsman, J., Miller, S., & Pfund, C. (2007). *Scientific teaching*: W.H. Freeman & Company in collaboration with Roberts & Company Publishers.

Hardré, P. L., & Burris, A. O. (2012). What contributes to teaching assistant development: Differential responses to key design features. *Instructional Science, 40*(1), 93–118.

Herrington, D. G., & Nakhleh, M. B. (2003). What defines effective chemistry laboratory instruction? Teaching assistant and student perspectives. *Journal of Chemical Education, 80*(10), 1197–1205.

Hollar, K., Carlson, V., & Spencer, P. (2000). $1 + 1 = 3$: Unanticipated benefits of a co-facilitation model for training teaching assistants. *Journal of Graduate Teaching Assistant Development, 7*(3), 173–181.

Huitt, W., & Hummel, J. (2003). Piaget's theory of cognitive development. *Educational psychology interactive, 3*(2).

Injaian, L., Smith, A. C., German Shipley, J., Marbach-Ad, G., & Fredericksen, B. (2011). Antiviral drug research proposal activity. *Journal of Microbiology & Biology Education, 12*, 18–28.

Kendall, K. D., Niemiller, M. L., Dittrich-Reed, D., Chick, L. D., WIlmoth, L., Milt, A., ... Schussler, E. E. (2013). Departments can develop teaching identities of graduate students. *CBE Life Sciences Education, 12*(12), 316–317.

Kolb, D. A. (1984). *Experiential learning: Experience as the source of learning and development* (Vol. 1). Englewood Cliffs, NJ: Prentice-Hall.

Lawrence, G. D. (2009). *People types & tiger stripes: Using psychological type to help students discover their unique potential* (4th ed.). Gainesville, FL: Center for Applications of Psychological Type.

Leman, T., & Acar, S. B. (2012). Jigsaw cooperative learning: Acid-base theories. *Chemistry Education Research and Practice, 13*(3), 307–313.

Lombardi, S. A., Hicks, R. E., Thompson, K. V., & Marbach-Ad, G. (2014). Are all hands-on activities equally effective? Effect of using plastic models, organ dissections, and virtual dissections on student learning and perceptions. *Advances in Physiology Education, 38*(1),80–86. doi:10.1152/advan.00154.2012.

Luft, J. A., Kurdziel, J. P., Roehrig, G. H., & Turner, J. (2004). Growing a garden without water: Graduate teaching assistants in introductory science laboratories at a doctoral/research university. *Journal of Research in Science Teaching, 41*(3), 211–233.

Luo, J., Bellows, L., & Grady, M. (2000). Classroom management issues for teaching assistants. *Research in Higher Education, 41*(3), 353–383.

Marbach-Ad, G. (2009). From misconceptions to concept inventories. *Focus on Microbiology Education, 15*(2), 4–6.

Marbach-Ad, G., McAdams, K., Benson, S., Briken, V., Cathcart, L., Chase, M., et al. (2010). A model for using a concept inventory as a tool for students' assessment and faculty professional development. *CBE Life Science Education, 9*(408–436).

Marbach-Ad, G., Rotbain, Y., & Stavy, R. (2005). Using a bead model to teach high-school molecular biology. *School Science Review, 87*, 39–52.

Marbach-Ad, G., Rotbain, Y., & Stavy, R. (2008). Using computer animation and illustration activities to improve high school students' achievement in molecular genetics. *Journal of Research in Science Teaching, 45*(3), 273–292. doi:10.1002/tea.20222.

Marbach-Ad, G., Schaefer, K. L., Kumi, B. C., Friedman, L. A., Thompson, K. V., & Doyle, M. P. (2012). Development and evaluation of a prep course for chemistry graduate teaching assistants at a research university. *Journal of Chemical Education, 89*(7), 865–872.

Marbach-Ad, G., Shields, P. A., Kent, B. W., Higgins, B., & Thompson, K. V. (2010). Team teaching of a prep course for graduate teaching assistants. *Studies in Graduate and Professional Students Development, 13*, 44–58.

Marbach-Ad, G., Schussler, E. E., Miller, K., Ferzli, M., & Read, Q. D. (2015, April). *Professional development for biology graduate teaching assistants: Status, challenges and needs*. Poster presented in the National Association for Research in Science Teaching [NARST] annual meeting, Chicago, IL.

Marbach-Ad, G., & Sokolove, P. G. (2000a). Can undergraduate biology students learn to ask higher level questions? *Journal of Research in Science Teaching, 37*(8), 854–870.

Marbach-Ad, G., & Sokolove, P. G. (2000b). Good science begins with good questions: Answering the need for high-level questions in science. *Journal of College Science Teaching, 30*(3),192–195.

Marbach-Ad, G., & Stavy, R. (2000). Students' cellular and molecular explanations of genetic phenomena. *Journal of Biological Education, 34*(4), 200–205.

Marbach-Ad, G., & Sokolove, P. G. (2001). The use of E-mail and in-class writing to facilitate student-instructor interaction in large-enrolment traditional and active learning classes. *Journal of Science Education and Technology, 11*, 109–119.

Marbach-Ad, G., Yarden, H., & Gershoni, J. M. (2007). Using the concept map technique as diagnostic and instructional tool in introductory cell biology to college freshmen. *Journal of Student Centered Learning, 3*(3), 163–177.

Marbach-Ad, G., Katz, P., & Thompson, K. V. (in press). A disciplinary teaching certificate program for science graduate students. *Journal of Teaching and Learning Centers*.

Mayer, R. E. (2002). Rote versus meaningful learning. *Theory Into Practice, 41,* 226–232.

McKeachie, W. J., & Svinicki, M. D. (2014). *McKeachie's teaching tips: Strategies, research, and theory for college and university teachers* (14th ed.). Belmont, CA: Wadsworth.

Micomonaco, J. P. (2011). *CIRTL: Impacting STEM education through graduate student professional development.* Paper presented at the 118th Annual American Society of Engineering Education (ASEE) Annual Conference and Exposition, Vancouver, British Columbia. Retrieved from tinyurl.com/8xcflgc

Novak, G. N., Patterson, E. T., Gavrin, A., & Christian, W. (1999). *Just-in-time teaching: Blending active learning and web technology.* Saddle River, NJ: Prentice Hall.

Novak, J. D., & Gowin, D. B. (1984). *Learning how to learn.* London, England: Cambridge University Press.

Nyquist, J. D., Manning, L., Wulff, D. H., Austin, A. E., Sprague, J., Fraser, P. K., . . . Woodford, B (1999). On the road to becoming a professor: The graduate student experience. *Change, 31*(3), 18–27.

Phillips, A. R., Robertson, A. L., Batzli, J., Harris, M., & Miller, S. (2008). Aligning goals, assessments, and activities: An approach to teaching PCR and gel electrophoresis. *CBE Life Sciences Education, 7*(1), 96–106. doi:10.1187/cbe.07-07-0052.

Poundstone, W. (1989). *Labyrinths of reason: Paradox, puzzles, and the frailty of knowledge.* New York, NY: Anchor.

Prieto, L. R., & Altmaier, E. M. (1994). The relationship of prior training and previous teaching experience to self-efficacy among graduate teaching assistants. *Research in Higher Education, 35*(4), 481–497.

Prieto, L. R., Yamokoski, C. A., & Meyers, S. A. (2007). Teaching assistant training and supervision: An examination of optimal delivery modes and skill emphases. *The Journal of Faculty Development, 21*(1), 33–43.

Rotbain, Y., Marbach-Ad, G., & Stavy, R. (2005). Understanding molecular genetics through a drawing-based activity. *Journal of Biological Education, 39*(4), 174–178.

Rotbain, Y., Marbach-Ad, G., & Stavy, R. (2006). Effect of bead and illustrations models on high school students' achievement in molecular genetics. *Journal of Research in Science Teaching, 43*(5), 500–529. doi:10.1002/tea.20144.

Rotbain, Y., Marbach-Ad, G., & Stavy, R. (2008). Using a computer animation to teach high school molecular biology. *Journal of Science Education and Technology, 17*(1), 49–58. doi:10.1007/s10956-007-9080-4.

Rotbain, Y., Stavy, R., & Marbach-Ad, G. (2008). The effect of different molecular models on high school students' conceptions of molecular genetics. *The Science Education Review, 7*(2), 59–64.

Rushin, J. W., De Saix, J., Lumsden, A., Streubel, D. P., Summers, G., & Bernson, C. (1997). Graduate teaching assistant training: A basis for improvement of college biology teaching and faculty development. *Journal of American Biology Teacher, 59*(2), 86–90.

Sauermann, H., & Roach, M. (2012). Science PhD career preferences: Levels, changes, and advisor encouragement. *PloS One, 7*(5), e36307.

Schonborn, K. J., & Anderson, T. R. (2008). Bridging the educational research-teaching practice gap – Conceptual understanding, part 2: Assessing and developing student knowledge. *Biochemistry and Molecular Biology Education, 36*(5), 372–379. doi:10.1002/Bmb.20230

Schussler, E. E., Torres, L. E., Rybczynski, S., Gerald, G. W., Sarkar, P., Shahi, D., . . . Osman, M. A. (2008). Transforming the teaching of science graduate students through reflection. *Journal of College Science Teaching, 38*(1), 32–36.

Seymour, E., & Hewitt, N. M. (1997). *Talking about leaving: Why undergraduates leave the sciences.* Boulder, CO: Westview Press.

Shulman, L. S. (1986). Those who understand: Knowledge growth in teaching. *Educational Researcher, 15*(2), 4–31.

Singh, S. (1997). *Fermat's Enigma: The epic quest to solve the world's greatest mathematical problem*. New York, NY: Walker and Company.

Sinnott, C., Marbach-Ad, G., Orgler, M., & Thompson, K. V. (2011). *Educational goals and insight from classroom experiences reported by graduating seniors in science majors*. Paper presented at the Annual Lilly Conference on College and University Teaching, Washington DC.

Smith, M. K. (2008). *Howard Gardner theory, multiple intelligences and education*. Retrieved March 15, 2014.

Sokolove, P. G., & Marbach-Ad, G. (1999). The benefits of out-of-class group study for improving student performance on exams: A comparison of outcomes in active-learning and traditional college biology classes. *Journal on Excellence in College Teaching, 10*, 49–68.

Stavy, R., Strauss, S., Orpaz, N., & Carmi, G. (1982). U-shaped behavioral growth in ratio comparisons. In R. Stavy (Ed.), *U-shaped behavioral growth* (pp. 11–36). Oxford, UK: Elsevier, INC.

Stavy, R., & Tirosh, D. (2000). *How students (mis-) understand science and mathematics: Intuitive rules*. New York, NY: Teachers College.

Stavy, R., & Wax, N. (1989). Children's conceptions of plants as living things. *Human Development, 32*(2), 88–94.

Tanner, K., & Allen, D. (2006). Approaches to biology teaching and learning: On integrating pedagogical training into the graduate experiences of future science faculty. *CBE-Life Sciences Education, 5*(1), 1–6.

Thijs, G. D., & van den Berg, E. (1993, August 1–4). *Cultural factors in the origin and remediation of alternative conceptions*. Paper presented at the Third International Seminar on Misconceptions and Educational Strategies in Science and Mathematics, Cornell University, Ithaca, NY.

Tirosh, D., Stavy, R., & Aboulafia, M. (1998). Is it possible to confine the application of the intuitive rule: Subdivision processes can always be repeated? *International Journal of Mathematical Education in Science and Technology, 29*(6), 813–825.

Wieman, C. (2007). Why not try a scientific approach to science education? *Change.*www.changemag.org/Archives/Back%20Issues/September-October%202007/index.html

Yarden, H., Marbach-Ad, G., & Gershoni, J. M. (2004). Using the concept map technique in teaching introductory cell biology to college freshmen. *Bioscene, 30*, 3–13.

Chapter 6
Evaluating the Effectiveness of a Teaching and Learning Center

Program evaluation plays an important role in all developmental stages of a teaching and learning center. This evaluation informs program development, provides feedback for program improvement, and documents the impact of a center for accountability purposes (Plank & Kalish, 2010). Our first step in creating our Teaching and Learning Center (TLC) for chemistry and biology faculty was to conduct a comprehensive needs assessment. We collected information about existing professional development initiatives in teaching and learning, plans and goals for future initiatives, and ongoing challenges within the departments. This initial needs assessment shaped the TLC's mission and planning. In addition, it has enabled us to tailor TLC programming to enhance existing initiatives and establish new initiatives based on our stakeholders' goals and needs. Ongoing assessment has been integrated into all aspects of TLC programming and services to provide feedback on five different attributes: participation, satisfaction, learning, application, and overall impact (Colbeck, 2003; Kirkpatrick, 1994). This iterative process has informed our program planning and helped us to refine our activities over time. In this chapter, we discuss the role of program evaluation for a teaching and learning center and illustrate with examples from our evaluations of the TLC.

The Role of Comprehensive Program Evaluation

In recent decades, researchers and practitioners have increasingly recognized that comprehensive evaluation serves a vital role in professional development programs (Austin & Sorcinelli, 2013; Austin, Sorcinelli, & McDaniels, 2007; Chen, Kelley, & Haggar, 2013; Fink, 2013; Guskey, 2000; Hines, 2009; Kucsera & Svinicki, 2010; Levinson-Rose & Menges, 1981; Plank & Kalish, 2010; Sorcinelli, Austin, Addy, & Beach, 2006; Wright, 2011). Over this time period, standards for rigorous evaluation have come to incorporate not just output-based evaluation—e.g., the

© Springer International Publishing Switzerland 2015   185
G. Marbach-Ad et al., *A Discipline-Based Teaching and Learning Center*,
DOI 10.1007/978-3-319-01652-8_6

number of trainings offered—but also outcome-based evaluation—e.g., the degree to which participants learn from and change their behaviors based on that training (Chen et al., 2013). However, conducting outcomes-based evaluation is complex and generally requires extensive time and expertise. Furthermore, most teaching and learning centers are based on the premise that change is a multi-step process, with proximal and distal outcomes, and the desired distal outcomes may be difficult to measure. Even when outcomes have been measured, it can be difficult to attribute causality to center activities and interventions.

We address this complexity by employing a multi-level evaluation plan that reflects our theory of action (Fink, 2013; Plank & Kalish, 2010). The ultimate goal of providing professional development to faculty members and graduate students is to impact undergraduate education. Therefore, we collect evaluation data from three populations: the faculty members and graduate students whom we serve, and the undergraduates that are the recipients of instruction. Collecting comparable, contemporaneous data from these three populations allows us to triangulate our findings and enables us to evaluate our program on multiple levels.

In the sections that follow, we provide recommendations for instituting a rigorous evaluation program. We first discuss the role of needs assessments in the initial stage of the evaluative process. We then describe the evaluation of our professional development program and its individual components.

Needs Assessment: A First Step

Our Teaching and Learning Center emerged from a growing consensus among College leadership regarding the importance and complexity of effective teaching. National reports (AAAS, 1990; Boyer Commission on Educating Undergraduates in the Research University, 1998; NRC, 2003; NSF, 1996) raised the salience of undergraduate teaching as a target of institutional reform. The growing body of literature on scientific teaching (Handelsman et al., 2004; Wieman, 2007) presented several models for reform, each of which would benefit from sustained and comprehensive professional development to support their implementation. As a key step in providing this professional development, our College leadership suggested the formation of an in-house program of professional development focused on improving instruction and thereby enhancing student learning. Once this decision had been made, the College secured start-up funding from a Howard Hughes Medical Institute Undergraduate Science Education Grant to create the TLC and hired a science education specialist to serve as its founding director.

With this infrastructure in place, the new director spearheaded a needs assessment process to understand the context in which the TLC would work and its role within this context. Such needs assessments, also known as context evaluations, serve multiple objectives that include defining needs, problems, assets, opportunities, and purposes within the environment in which services will be provided

(Stufflebeam, Madaus, & Kellaghan, 2000). The needs assessment should first identify *needs* in terms of desired resources or services, as well as the *problems* or impediments to meeting those needs. The next step is to determine relevant *assets* in terms of expertise, services, and resources, which can be leveraged to create *opportunities* to meet needs and overcome problems. This process should result in the refinement of a *purpose* or mission that defines and delineates what the program can and should accomplish given the specific context in which it operates. Ongoing needs assessment constitutes an important component of continuous program improvement efforts and should continue throughout the lifespan of the program (Stufflebeam et al., 2000).

In terms of faculty development programs, needs assessments should include multiple stakeholders that reflect the context of the institution and the focus of the program under consideration (Travis, Hursh, Lankewicz, & Tang, 1996). Including all relevant stakeholders in this process provides the most complete representation of local needs, problems, and assets. In our context, this involved engaging key personnel at the university, college, and departmental levels. Through the meaningful involvement of a broad range of stakeholders, needs assessments can also serve to build a network of "allies" who will support the center's future work (Sorcinelli, 2002).

Stakeholders and Their Input into the TLC Needs Assessment

Our needs assessment involved input from a representative group of the TLC's stakeholders, as well as a review of existing resources (Fig. 6.1).

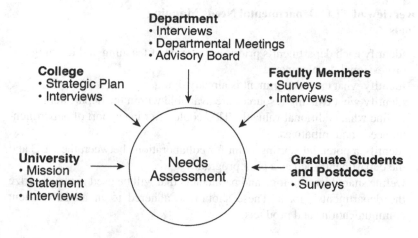

Fig. 6.1 Sources of input for TLC needs assessment

University Level: At the university level, we reviewed the University's Mission Statement, which outlines the institutional identity, capabilities, and objectives and outcomes. Our institution's first stated objective was to "continue to elevate the quality ... of undergraduate education" (University of Maryland Office of the Provost, 2011), which was in strong agreement with the TLC's proposed mission.

To better understand the scope of existing initiatives, we also interviewed the heads of campus-wide units that focus on teaching and learning, such as the Center for Teaching Excellence (CTE, which was situated within the campus Office of Undergraduate Studies) and the Office of Learning Technologies (which was part of the campus Office of Information Technology). One of our primary objectives in these interviews was to determine how our TLC could complement, but not supplant, the work of these centers.

College Level: Our needs assessment coincided with the College's development of a new five-year strategic plan. The TLC's mission was intertwined with the goals and objectives outlined in the strategic plan, which emphasized the importance of undergraduate education. The dean, as well as associate and assistant deans, met with the TLC director and provided feedback on the role that the TLC could play in addressing College needs and objectives. They also suggested key personnel who should be consulted during the course of the needs assessment.

Department Level: Given our discipline-specific focus and our broad goal of providing professional development tied to pedagogical content knowledge (PCK), our needs assessment at the departmental level was particularly extensive. Our emphasis on the department reflects recent research indicating that the department is the most "productive unit of change" for reform of undergraduate education (Gibbs, Knapper, & Piccinin, 2008; Quardokus & Henderson, 2014; Wieman, Perkins, & Gilbert, 2010).

Overview of TLC Departmental Needs Mapping
Goals

- Identify each department's priorities in terms of teaching and learning and professional development.
- Identify what each department is already doing.
- Identify what existing resources are available within the department.
- Define what additional value the TLC could offer in support of departmental needs and initiatives.
- Identify a potential starting point for collaborations between the TLC and the department, such as a pilot program.
- Define shared terminology and parameters that will be used to characterize the departments' goals. These should be adhered to in all subsequent communications and products.

(continued)

Gathering Information about the Departments

- Prior to the meeting, gather the following information (to make sure that we focus on the department's specific needs and not on generic needs):

  - The department vision, goals, organization structure, main activities, and relevant key personnel
  - Background about departmental subunits and the people with whom we are going to meet

- Send a short survey to departmental leadership to identify:

  - Gaps between existing and desired activities
  - Key personnel to speak with and involve in future initiatives

- Assess the current state of professional development activities for:

  - New faculty (e.g., mentoring and acculturation)
  - Current faculty (e.g., ongoing support, assistance with teaching quandaries)
  - Graduate students

We interviewed the department chairs and other personnel with key roles in the delivery of undergraduate instruction, such as the directors of undergraduate and graduate studies (see Guiding Questions for Departmental Meetings in the Implementation Guide at the end of this chapter). Prior to the interviews, we asked departmental chairs to collect information for discussion in the interview, and we provided them with a template for mapping current initiatives as well as needs and opportunities for the future. The Mapping Template includes sections on initiatives for undergraduates, graduate students, faculty members, and outreach. We recommended that department chairs invite key personnel involved in these initiatives to participate in the interview.

Based on interview transcripts and information from follow-up discussions, TLC staff compiled a comprehensive summary of all the departmental teaching and outreach activities, and added missing information to the maps. Each department chair reviewed the information in the updated Map for accuracy. Table 6.1 shows excerpts from the tables of different departments to illustrate the range and scope of activities across the College.

After completing the Departmental Needs Mapping Templates, TLC staff gave presentations at department meetings about the new Center, its mission, and the availability of professional development activities. In each meeting, the overview of planned professional development activities reflected each department's needs and goals as identified in the needs mapping process. We asked all faculty members to provide feedback and suggestions, particularly in terms of how to better

Table 6.1 Sample needs mapping table excerpts

Type and focus of program	Recent Initiatives	Contact person	Needs
Undergraduate students (Biology)			
Course reform	Cell Biology and Physiology BSCI 330 (HHMI)—pedagogy and curricular reform	Course instructors	Mechanism for coordinating sequences of courses
	Organismal Biology (BSCI 207)—pedagogy and curricular reform		
Teaching/mentoring opportunities for science majors	BSCI 329—course using undergraduate teaching assistants		
	Prep program for undergraduate teaching assistants	Faculty coordinator	Assistance with science education topics
Other	Number of Departmental Honors degrees has tripled		
Graduate Students (Chemistry and Biochemistry)			
Prep course for new GTAs	Mandatory prep course for new Chemistry and Biochemistry GTAs	Course instructors	UTLP needs to be pushed more
Opportunities for professional development and dissemination	GAANN grants to support graduate students in areas of national need	Faculty coordinator	Training in teaching and mentoring
Other	Participation in teaching and learning conferences		Additional funding
Faculty (Cell Biology and Molecular Genetics)			
Acculturation of new faculty	Every new faculty member is assigned a mentor	Chair	
Faculty professional development in teaching	Host Pathogen Interaction FLC	FLC Coordinator	Assistance with grant writing and data analysis
Travel and other grant opportunities for teaching and learning	Grant support for implementing blended learning in BSCI 410 upper-level course	Course instructors	Assistance with grant implementation
Other			

tailor professional development activities to their department and individual needs, interests, and goals.

For purposes of maintaining ongoing two-way communication between the TLC and its constituents, each department appointed a well-respected, tenure-track faculty member to serve on the TLC Advisory Board. The Advisory Board meets annually or as needed to review needs, provide feedback on existing programming, and make recommendations for new professional development initiatives.

Faculty Members: In addition to seeking faculty member input at department meetings, we have also used interviews and surveys to obtain a more comprehensive understanding of faculty members' needs and priorities. In 2009, we distributed the pilot Survey of Teaching Beliefs and Practices (STEP-pilot)[1] to all faculty members in the departments we serve. The STEP-pilot included 28 items that encompassed (a) the skills that faculty members believe are most important for students to acquire over the course of their undergraduate education, (b) faculty members' beliefs regarding the most important teaching approaches to use, (c) faculty members' reported use of specific teaching approaches, and (d) faculty members' professional development needs and challenges with regard to teaching and learning. Approximately 30 % of faculty members responded to the pilot survey.

Since the 2009 pilot, the STEP has gone through comprehensive validation and multiple iterations (Marbach-Ad, Shaefer-Ziemer, Orgler, & Thompson, 2014); a validated version of the final survey can be found online at www.cmns-tlc.umd.edu/tlcmeasurementtools. We continue to collect faculty feedback by periodically administering the STEP. Faculty feedback from this survey is a component of our ongoing needs assessment. TLC staff and departmental leadership review this feedback periodically to determine how TLC professional development offerings can be improved and/or adapted to better meet faculty needs.

STEP Item: What Kinds of Professional Development Programs Would Help You with Your Teaching Responsibilities?

Feedback on teaching from an impartial observer.

Limited in-class visits and accompanying tutoring in alternative and novel practices.

Observing successful teachers teach.

Coordination among classes . . . I will admit that I'm really not sure what knowledge instructors in subsequent classes expect students to come away from my class with.

(continued)

---

[1]The design of the STEP-pilot was informed by some items from the Higher Education Research Institute's (HERI) faculty survey (Hurtado, Eagan, Pryor, Whang, & Tran, 2012). The HERI survey is available at www.heri.ucla.edu.

> *A retreat type thing to really focus on changing the structure of class meetings to be more "flipped." I think this is something that needs to be done in one concentrated period of time, not bit by bit.*
>
> *Small workshops, summer programs.*
>
> –Select faculty member responses

We also gathered information through targeted interviews. These interviews, which comprised a 3-year longitudinal study, focused specifically on the needs of faculty members who were new to the institution. The interviews are described in detail in Chap. 3 of this volume, as well as in Marbach-Ad, Schaefer Ziemer, Thompson, & Orgler (2013).

Graduate Students: At our institution, graduate students play a key role in providing undergraduate instruction through graduate teaching assistantships. To assess their professional development needs, we administered a survey similar to the STEP-pilot. The survey consisted of 22 items, many of which paralleled questions on the faculty STEP survey.

In addition to the formal mechanisms for stakeholder input described above, our understanding of the needs, problems, assets, and opportunities for the TLC was informed by the TLC staff's accumulated knowledge of our local context and informal conversations with administrators, faculty, and graduate students.

Final Steps of the Needs Assessment Process

The final component of a needs assessment as defined by Stufflebeam, Madaus, and Kellaghan (2000) is the refinement of the program's purpose and the delineation of the goals to be accomplished. In this step, we considered both recommendations from the literature and the findings from our needs assessment process. Robertson (2010) suggested that teaching and learning centers can focus on some combination of the following four areas for improving undergraduate education:

1. Instructional development: helping those who teach to learn to do it ever more effectively;
2. Faculty development: helping faculty members with all aspects of faculty work across their careers;
3. Curriculum development: facilitating instructional design (integrated learning goals, activities, and assessment) in the contexts of course units up to whole programs such as general education or degree programs; and
4. Organizational development: helping the institution to develop as an intentional learning organization in order to enhance strategic institutional effectiveness, for example, through a new chairperson and staff development (p. 39).

Our TLC addresses the first, third, and part of the fourth of Robertson's (2010) suggested areas. We are strongly committed to helping faculty members and graduate students develop their expertise as teachers, but do not focus on faculty responsibilities outside of teaching. We also work with individual faculty members and groups of faculty members in curriculum development. In terms of organizational development, we seek to foster an intentional learning community that strongly values effective teaching and offers its members ample opportunities for development in this area. Our organizational development focus does not include specific and deliberate work on policies or structures that impact the organizational environment around teaching (e.g., workloads, tenure and promotion considerations). However, we recognize that these environmental and structural elements are important components of facilitating change in higher education (Henderson, Beach, & Finkelstein, 2011), and we collaborate with other institutional stakeholders to foster this environment when opportunities arise. This delineation of purposes informed our mission statement and offers us a set of goals that we can reasonably expect to accomplish. Based on this broad definition, our program includes a menu of activities that are tailored to the population we serve and the goals we seek to achieve.

Our program of activities is targeted and differentiated. Some activities are intended for the entire College and others for subsets of this population. For example, we offer workshops that cover a range of topics that are relevant for faculty members, postdocs, and graduate students from all of the departments we serve (see Chap. 2 for more details). Some needs, however, are more appropriately addressed through more narrowly targeted professional development. We offer workshops specifically for new instructors (Chap. 3), luncheons for non-tenure-track faculty with instructional responsibilities (Chap. 4), and multiple opportunities for graduate students to develop their knowledge and skills in teaching (Chap. 5). We also offer highly personalized consultations to individuals and groups of faculty members (Chap. 4). While most of this professional development is optional, some components are required. All incoming faculty members are required to attend a short welcome workshop, and all new graduate assistants must complete a six-week preparatory course. In addition to differences in focus, our professional development opportunities vary in terms of their length. We offer short, discrete activities as well as long-term professional development opportunities. This variation enables us to accommodate the different needs, interests, capacities, and goals of the populations we serve.

Ongoing Evaluation: Measuring Five Levels of Program Impact

With the needs assessment completed and a new slate of activities underway, our evaluation focus shifted to the impact of our programs. We employ a model for evaluating professional development programs in teaching and learning

(Colbeck, 2003; Guskey, 2000; Kirkpatrick, 1998) that includes the following five levels of program evaluation:

1. Participation
2. Satisfaction
3. Learning
4. Application
5. Impact

Table 6.2 summarizes the data collection methods, assessment tools, and what can be learned from each level of program evaluation. On the pages that follow, we describe in detail each of these levels of evaluation and show how they contribute to comprehensive teaching and learning center evaluation.

Level 1: Participation

We maintain detailed records of all TLC program activities to allow us to analyze who participates in these activities, their motivation for participating, and the types of activities in which they participate. In the early years of the TLC, these records also provided useful insight into the success of our implementation plan and they continue to be useful in tailoring future programming to the evolving needs of our constituents.

How Do Participants Know About the TLC and Its Activities?

Potential participants in TLC activities can only take part if they are aware of these activities. To ensure broad awareness, we publicize our programs through multiple channels, including announcements at faculty meetings, messages sent to departmental email lists, web postings, postings to the college and university events calendars, and in casual conversations with faculty members and graduate students. To understand how well these marketing methods work and if they succeed in reaching our target population, we include a question on this topic each time we administer the STEP.

Recent results indicate that faculty members learned about the TLC through various mechanisms, and that fewer than 5 % of survey respondents were unaware of the TLC prior to taking the survey (Table 6.3). This has reinforced our belief in the need for multiple methods of communicating with our target audience.

Who Participates and in *What* Do They Participate?

Our TLC serves three life sciences departments (biology, entomology, and cell biology and molecular genetics [CBMG]) and one chemistry department (chemistry

Table 6.2 Outline of the five levels of program evaluation for a teaching and learning center

Level of evaluation	Data collection methods and assessment tools	What can be learned from the data collected
Participation	Detailed attendance records, with demographic information (e.g., gender, faculty rank, student major)	Who is participating in TLC activities, and for which demographic groups are more concerted recruitment efforts necessary?
	Surveys and interviews that probe motivation to participate	How can TLC programs attract a greater diversity of participants?
	Surveys and interviews to inquire how participants learned of TLC programs	How can the TLC better publicize its activities?
Satisfaction	Surveys and interviews to gather feedback on participant expectations	How successful was the activity or program in achieving its goals (from the perspective of both the organizers and the participants)?
		How can we modify the activities to better satisfy the expectations of the target audience?
Learning	Surveys and interviews of faculty and GTAs regarding their beliefs about teaching	How successfully has the TLC raised faculty and GTA awareness of the variety of evidence-based teaching approaches?
	Surveys and interviews of faculty and GTAs regarding their understanding of how to employ evidence-based teaching approaches	How successfully has the TLC assisted faculty and GTAs in employing evidence-based teaching approaches?
Application	Surveys and interviews asking faculty and GTAs about their use of different teaching practices	How does instructors' awareness of evidence-based teaching approaches affect their teaching?
	Classroom observations using an observation rubric	What are the challenges of implementing evidence-based teaching approaches?
		How can the TLC support instructors trying to overcome these challenges?
Impact	Surveys of students about teaching and assessment methods that they experienced during their degree programs, and their views about these methods	How well do undergraduate student perspectives on their overall educational experience correspond to the instructor perspectives?
		How can we prepare students to be more accepting of student-centered teaching approaches and able to learn effectively in this type of environment?

Table 6.3 Faculty members' responses to the 2011 STEP Survey question: How did you learn about the TLC? (Check all that apply)

Departmental meeting	38.2 %
TLC website	5.9 %
Email from TLC director	58.8 %
Colleague told me about TLC	30.9 %
New faculty workshops/interviews	19.1 %
Other (e.g., know TLC director, have worked with TLC repeatedly)	14.7 %
Did not know TLC exists	4.4 %

N = 68 survey respondents

and biochemistry) as well as college-level administrators and faculty. In addition to these main constituents, our seminars and workshops draw participants from across the campus and sometimes beyond. At each professional development activity, participants sign in with their name, affiliation, and position within the university.

In tracking participation in TLC activities, we find it valuable to not only track numbers of participants, but also to disaggregate participants by key demographic factors such as discipline, departmental affiliation, and position or role within the university. This helps us to determine whether we are successful in attracting our target populations. Additionally, we use this information to plan future programming to attract previously underrepresented populations. For example, when we noted that participation from one department was consistently disproportionately low, we sought feedback from the department chair and other key departmental personnel to plan future activities that better complemented their interests and current initiatives.

While our primary focus is on the chemistry and biology departments, our workshops and seminars are advertised to the entire campus community and open to all. We often have participants from related scientific fields (e.g., physics, mathematics), the College of Education, and other nearby universities. This philosophy of inclusion has strengthened our interdisciplinary curricular initiatives and facilitated the sharing of ideas among a broader population with a shared interest in science education.

Tracking the number of activities in which individuals participate enables us to identify faculty members with strong interests in teaching and learning. Documentation of individual participation also allows a fuller understanding of the context in which individual faculty members seek assistance. For example, when faculty members seek consultation from the TLC, we review which workshops they have attended so that we can build on their prior knowledge. We also draw from these attendance records in cases where faculty members undergoing promotion and tenure review ask the TLC for documentation of their professional development activities. Finally, we use information on the cumulative attendance of individual faculty members in later stages of our program evaluation process to measure outcomes related to the content of the training (e.g., are participants who attended trainings on different active learning techniques using these techniques in the classroom?).

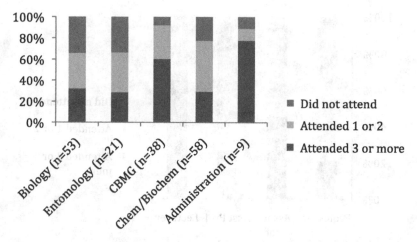

Fig. 6.2 Participation in TLC professional development activities by departmental affiliation

Program Evaluation Using Disaggregated Participation Data

In counting professional development activities, we consider each workshop or seminar a unique activity. We count participation in consultations or communities as a single activity even though these typically consist of multiple related meetings. This method results in a conservative count of cumulative participation. For easy visualization and interpretation, we have categorized participants' cumulative participation into three categories: did not attend any TLC activities, attended one or two activities, and attended three or more activities.

Figure 6.2 shows cumulative attendance by departmental affiliation. Between 2007 and 2013, 77 % of the faculty members in the departments we serve attended at least one TLC professional development activity. The department of CBMG had the highest proportion of faculty participating and the highest rate of participation. We attribute this to the long-standing Host-Pathogen Interaction (HPI) community, which has engendered strong interest in teaching and learning within that department. The data also demonstrate the strong support of the college administration for TLC activities. This group is small but important. It includes the assistant dean, associate dean, and other directors whose work involves undergraduate education. We have found over the years that maintaining administrator involvement is integral to ensuring sustained faculty engagement because it demonstrates institutional commitment to high quality undergraduate education.

We also disaggregated cumulative attendance by faculty member rank (professor, associate professor, assistant professor, or non-tenure-track lecturer). As Fig. 6.3 demonstrates, participation varied by rank, with the lecturer rank having the highest proportion of faculty participating and the highest rate of participation. Their disproportionately high participation rate clearly reflects their specific focus on teaching, but it is also indicative of college leadership's desire to provide ample professional development opportunities for this group.

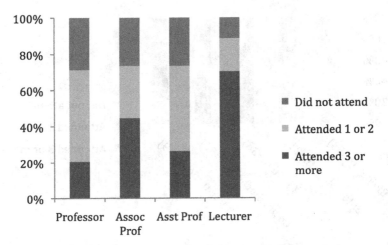

Fig. 6.3 Participation in TLC professional development activities by faculty rank

Tenured and tenure-track faculty also showed high rates of engagement, despite their additional scientific research responsibilities. More than 70 % of tenure-track faculty members participated in at least one TLC activity between 2007 and 2013, with associate professors showing the greatest tendency to participate in three or more TLC activities during this time. We are encouraged by these statistics as they indicate that, rather than being satisfied with the educational status quo, faculty generally are interested in learning about more effective teaching approaches and willing to participate in institutional efforts to improve undergraduate education.

While we do not calculate individual attendance percentages for graduate students due to their large numbers and relatively high turnover rates, we track their cumulative participation in a manner similar to how we track faculty participation. Between 2007 through 2013, more than 80 graduate students attended at least one of our open workshops and seminars (i.e., those that were targeted to faculty members and administrators in addition to graduate students, as opposed to those offered exclusively to graduate students). Additionally, since the six-week preparatory course became mandatory, approximately 70 new graduate teaching assistants have completed the course each year. Enrollment in the more advanced two-credit graduate course on teaching and learning averages about 12 students per year. Twenty-three graduate students have participated in the TLC's branch of the University Teaching and Learning Program since its inception in 2010 (UTLP; for a detailed description see Chap. 5).

We are heartened by the strong participation of graduate students in all levels of TLC programming, as they represent the future of academia. One of the primary missions of our TLC is to imbed substantive training in teaching science into graduate student education. As a result, evaluating the extent to which graduate students take advantage of TLC activities, especially those that are noncompulsory, is a high priority.

Table 6.4 Faculty responses to the 2011 STEP Survey question: Rate the importance of the following to your interest in professional development in teaching (response options: not important, slightly important, fairly important, important, very important)

Motivating factor	Important or very important (%)
Personal desire to improve professional skills	87
Helping prepare the next generation of young scientists	86
Ensuring that all students are scientifically literate	81
Support of the dean/department chair/undergraduate program director	76
Belonging to a teaching community	67
National efforts to promote science teaching	64
Information/assistance provided by the TLC	58
The College/University effort (i.e., strategic plan)	57
Promotion and tenure	52

N = 68 survey respondents

Why Do Individuals Participate?

It is important to collect data on motivation for participating in professional development related to teaching, both generally and with respect to specific activities (Colbeck, 2003). Motivation may be extrinsic (e.g., driven by institutional requirements, recommendations, or potential rewards), intrinsic (e.g., based on a faculty members' own desire to improve teaching), or a combination of the two. Understanding what individuals hope to achieve through professional development helps TLC staff to develop programming appropriate to their needs, and it can identify levers for increasing faculty and graduate student participation rates.

We periodically collect such information through the STEP survey (Table 6.4 illustrates survey results from 2011). STEP results indicate that many faculty members are motivated intrinsically, by their desire to improve their teaching or to better prepare future generations of scientists and citizens. Faculty members also highlighted the importance of some extrinsic sources of motivation. For example, the department plays an important role in motivating faculty members to participate in professional development. Based on this knowledge, the TLC works regularly with departmental leadership to ensure that TLC programming is in alignment with departmental goals, thereby ensuring continued departmental support for faculty participation in TLC activities.

We also collect data on why participants engaged in specific professional development activities. We have found it particularly valuable to understand what motivates participants to engage in labor-intensive professional development such as faculty learning communities (FLCs; described in Chap. 4) or the University Teaching and Learning Program (UTLP) for graduate students (Chap. 5), given the time involved in participating in these long-term professional development activities.

In our evaluation of the UTLP, for example, we gathered motivational information through semi-structured interviews of five graduate students who had finished the program and two graduate students who started the program but opted not to complete it. One of the interview questions explored students' motivation for participating in the program. Three major themes emerged:

1. Sharing their enthusiasm for science with students

   I love working with students. I love helping them understand and figure things out on their own, and watching them become more independent. I want to teach undergraduate science and especially research. I want to teach students how to be independent researchers. I can't seem to get away from how much I like teaching.

2. Acquiring skills for future teaching jobs and getting a teaching certificate noted on their diploma

   I knew about the notation on the transcript, which was important. Even though I was interested in the teaching and learning, and I was going to the seminars about teaching and learning, that's not reflected anywhere in my transcript unless I did this program. That was important to me.

3. Developing communication skills that are useful for teaching as well as other career pathways

   It teaches you many different things that can be used in other places too. It's not limited to teaching science only. I think what was covered in this program was beyond teaching.

Participants frequently mentioned a specific interest in teaching. In some cases, this interest was ignited by their parents and other family members who worked in the field of education; in other cases, it grew out of the participants' prior teaching experience. UTLP participants saw the program as a way to engage deeply in teaching and learning, in a way that was complementary to the training in science research that their graduate programs provide.

Level 2: Satisfaction

The second level of evaluation involves assessing satisfaction with the program and its constituent activities. We evaluate satisfaction from the perspective of both program developers and program participants. From the perspective of program developers, we look at whether the program achieved predetermined goals (Colbeck, 2003). For program participants, we obtain reaction data (Kirkpatrick, 1998). Reaction data are generally gathered through evaluation surveys and reflect participants' like or dislike for a program component or programming as a whole (Boyle & Crosby, 1997). These data provide program developers with valuable feedback that can be used to improve the program to better meet participants' needs and expectations (Boyle & Crosby, 1997).

Assessing the Achievement of Programmatic Goals

The TLC has an overarching goal of improving instruction in our undergraduate biology and chemistry departments. In pursuit of this goal, we offer a variety of professional development activities, each of which has its own individual objectives. Our evaluation of the degree to which each type of professional development activity met its objectives depends on whether the activity is short-term or long-term. For short-term activities, including seminars and workshops, program developers have very specific information that they intend to convey. After each activity of this sort, program staff involved in program planning and delivery conduct a debriefing to discuss how well the activity satisfied its specific goals and how they might adjust programming to better satisfy any goals that were not completely achieved.

For long-term activities, such as faculty learning communities (FLCs), our evaluation includes reviews of meeting agendas and obtaining observational data at FLC meetings. These observations are conducted by either TLC staff or, in the case of grant-funded initiatives, external evaluators. These observations utilize a formal protocol that guides the observer through different measures that characterize the community meeting, such as the physical environment, degree to which all participants contribute to the discussion, type and quality of meeting leadership, and whether the meeting enabled progress toward clearly established goals. After observing an FLC meeting, the observer meets with the FLC leaders to discuss the findings. To illustrate the utility of this type of evaluation, we have included in the Implementation Guide an observation protocol completed by an external evaluator.

Assessing Participant Satisfaction

To measure participant satisfaction, we generally request that participants complete a brief evaluation survey at the end of a program activity. Such evaluation surveys vary by the type of activity. Seminar and workshop evaluation surveys, for example, tend to be short and usually include Likert-style questions that probe participants' reactions to different components of the workshop (e.g., presentation, handouts or materials, group activities, and overall usefulness). Surveys also include open-ended questions about what participants found to be most useful about the workshop and their suggestions for improving future iterations of the workshop.

Graduate student preparatory courses are among the more complex and extensive of our professional development activities; therefore we assess participant satisfaction through a series of more detailed surveys. As described in Chap. 5, students complete a short pre-course survey about their expectations and goals for the course, as well as a mid-course evaluation and an end-of-course evaluation. Collectively, responses to these surveys provide rich information for evaluating whether the course satisfied student needs and expectations. Course instructors use the pre- and mid-course feedback to plan for subsequent course meetings and use all three sets of information to plan for future years.

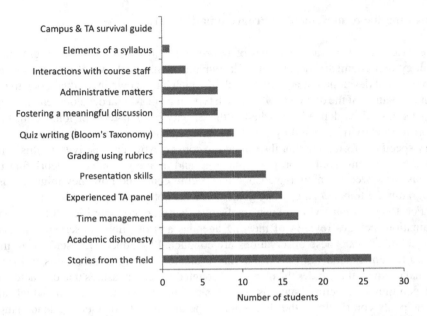

Fig. 6.4 Frequency with which GTA prep course topics were rated as most useful (N = 43 students. Each student could select three topics)

Program Evaluation Using a Satisfaction Survey

The end-of-semester evaluation survey from the first offering of the 6-week preparatory course for new biology GTAs in 2007 provided information about students' satisfaction with the overall course and its components. Survey results demonstrated that students generally found the course to be informative (80 % of course participants), interesting (56 %), and useful (56 %); while a very small minority found it to be boring (9 %), confusing (2 %), or a waste of time (2 %). When asked to identify the three most useful topics covered in the course, students highlighted the value of 'Stories from the field,' dealing with academic dishonesty, and effectively managing their time (see Fig. 6.4 for all responses; Marbach-Ad, Shields, Kent, Higgins, & Thompson, (2010) provides a more comprehensive description of the course and its evaluation).

> *Stories from the field and the experienced TA panel brought forth issues that are faced by TAs day in and day out. This 'pre-warned' us of the things to come.*
>
> *Stories from the field are just a good way to encourage talking and discussion, and encourage us to think while we get to laugh.*

(continued)

> *The discussion on academic dishonesty was useful as it provided me with a clear guideline to follow in case I detect a case amongst my students.*
>
> *My students had questions about how to do well, and time management was good advice to pass on.*
>
> –Student feedback from evaluation surveys on the three most useful topics in the GTA prep course

We generally take an informal approach to collecting feedback on participants' satisfaction with long-term activities such as FLCs and consultations. These activities involve regular interaction between participants and TLC staff. In these interactions, we periodically inquire about the degree to which the professional development satisfies their expectations. We also regularly seek their feedback about improvements or additional supports that they would find beneficial.

Level 3: Learning

This level of TLC program evaluation addresses what participants learn through our professional development activities. We measure the degree to which participants' beliefs about the knowledge and skills that students should acquire during their undergraduate studies align with the recommendations voiced in recent national reports (AAAS, 2011; PCAST, 2012). We also seek to measure their awareness of the teaching practices they can employ to instill this knowledge and these skills in their students. In order for instructors to employ appropriate and effective teaching practices, they generally need to progress through three learning stages:

1. Knowledge of evidence-based teaching practices,
2. Belief in the effectiveness of these practices in enhancing student learning, and
3. An understanding of how to employ these practices.

Stage 1: Knowledge of Evidence-Based Teaching Practices

In the evaluation of the first learning stage, we measure knowledge related to specific professional development activities. For example, after a workshop on blended learning, we may ask if participants left the workshop with a better understanding of what constitutes blended learning and different ways in which blended learning can be implemented. We also address this learning stage indirectly in our periodic faculty surveys. For example, we ask faculty members and graduate students to describe their teaching philosophy, which can reveal their knowledge of evidence-based teaching and learning practices. A well-developed teaching philosophy will describe skills that students should acquire as well as teaching practices that will support students in developing those skills.

Stage 2: Belief in the Effectiveness of Evidence-Based Teaching Practices in Enhancing Student Learning

We evaluate this learning stage through the periodic STEP survey, which asks faculty members and graduate students to rate the importance of various teaching practices (e.g., group work, inquiry-based learning, and scientific writing) and educational goals (e.g., the ability to work effectively in groups, understanding the dynamic nature of science, problem-solving capacity, and evidence-based decision-making). The STEP asks respondents to rate the importance of both traditional and evidence-based teaching practices on a five-point scale. In our analysis, we disaggregate responses by population characteristics to better understand all members of our multi-departmental community.

Program Evaluation Using a Survey of Beliefs in the Importance of Educational Goals

The results from our 2011 STEP survey highlight similarities and differences in faculty member and graduate student response patterns regarding educational goals (see Fig. 6.5). The two populations were similar in that they generally rated nationally recommended goals (i.e., understand how science applies to everyday life, understand the dynamic nature of science, and scientific writing) more highly than goals related to memorization of content and procedures (i.e., remember formulas, structures, and procedures and memorizing basic facts). However, the

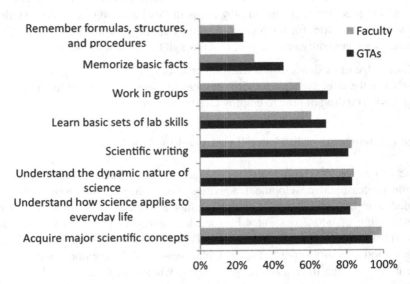

Fig. 6.5 GTA (N = 99) and faculty member (N = 71) ratings of the importance of specific educational goals for undergraduates. Percentages reflect combined categories 4 (important) and 5 (very important)

two populations differed in their ratings of the importance of working in groups; only 55 % of faculty members rated this as an *important* or *very important* skill, compared to 70 % of GTAs.

We found it counterintuitive that faculty members would not rate group work as an important educational goal, particularly because scientific research is often collaborative in nature. We suspect that the relatively low ratings for group work may stem from difficulties faculty members have experienced in implementing group work in the classroom. Many of our faculty members teach large classes, with enrollments of 200–300 students, which makes it particularly difficult to manage students working on semi-independent group activities. Additionally, many faculty members have received negative feedback from students about such activities, especially when the group work contributes to the course grade. For a more detailed analysis of these and other findings from the 2011 surveys, see Marbach-Ad et al. (2014, in press).

Stage 3: An Understanding of How to Employ Evidence-Based Teaching Practices

This learning stage relates to faculty member and graduate student capacity to employ evidence-based teaching practices. This stage can be measured through survey questions that probe their confidence with using various teaching practices. To reduce the burden on respondents, we do not currently include this type of question in our surveys, as the surveys are already quite lengthy. Other programs may wish to include this type of question in addition to or instead of other questions, based on their program goals.

While we do not systematically collect information about confidence in using various teaching approaches, we get a general sense of this by attending FLC meetings and through our consultations with faculty members. In these settings, we work closely with individuals and groups, and much of our support focuses on building their confidence and expertise in the use of specific teaching strategies. When faculty members do not exhibit strong levels of confidence in their ability to employ the teaching approach, we often provide them with literature that illustrates the successful use of the approach in question, recommend videos that demonstrate the use of the techniques, connect them with a faculty member who is already using the technique, and/or observe their classes and provide feedback.

Level 4: Application

In the fourth level of evaluation, we assess the degree to which faculty members and graduate students use evidence-based teaching practices in their classrooms. We have three means of measuring this:

1. Faculty self-reports of the teaching practices they used (via STEP survey)

2. Student reports of teaching practices they experienced (via exit survey of graduating seniors)
3. Classroom observations

These mechanisms provide complementary evaluative information that assists us in understanding the impact of specific interventions, and allows us to document broader patterns of institutional change.

Faculty Self-Reports of the Teaching Practices They Used

Our periodic STEP surveys include items that ask faculty members to report the teaching practices they used in an undergraduate course of their choosing. We request that, if they teach multiple courses, they direct their responses to a high-enrollment and/or lecture course, because these courses generally present more challenges in the implementation of evidence-based teaching practices. In a series of related items, we ask faculty members to indicate the frequency with which they employ various teaching and assessment practices (see STEP at www.cmns-tlc.umd. edu/tlcmeasurementtools for a complete list of practices included in the survey).

We analyze faculty self-reports in multiple ways. For example, we look at the association between the educational goals that faculty members rate as important and their reported use of instructional techniques that can help students achieve these educational goals (Marbach-Ad et al., 2014, in press). This is of interest to us because our previous surveys have shown a sizeable gap between faculty beliefs in the value of active learning approaches and their use of these approaches in the classroom (Marbach-Ad, Schaefer, & Thompson, 2012). We now aim to better understand the types of supports that are effective in moving faculty members from mere appreciation to confident utilization of these approaches.

Program Evaluation Using a Survey of Instructional Practices Used

We promote FLCs as a way of supporting the widespread adoption of evidence-based teaching approaches. To evaluate whether this strategy is having the desired effect, we disaggregate STEP survey data on instructional practices used according to whether or not the faculty members belong to an FLC. Table 6.5 shows faculty self-reports of their use of various teaching approaches from the 2011 STEP survey. Results indicated that traditional teaching approaches (such as extensive lecturing and answering questions from individual students during class) were widely and frequently used, independent of whether a faculty member belonged to an FLC. However, FLC participants reported significantly higher use of several evidence-based teaching approaches than non-FLC participants. These approaches included group work during class time, group work outside of class time, out-of-class discussions, debates during class time; and the use of games, simulations, and/or role-play.

Table 6.5 Faculty self-reported frequency of use of various classroom teaching approaches

Teaching approaches	Frequency of use	
	FLC	Non-FLC
Answering questions from students in class	4.53	4.67
Extensive lecturing	4.42	4.57
Asking students to interpret graphical information	3.60	3.20
Class discussions	3.55	3.27
Communicating course goals and objectives	3.48	3.50
Multimedia instruction	3.00	2.67
Group work outside of class time*	2.90	1.97
Group work during class time*	2.81	2.00
Real-life problems	2.55	2.43
Out-of-class discussions*	2.48	1.62
Personal Response System	2.35	1.70
Debates in class*	2.32	1.70
Online modules with immediate feedback	1.87	1.40
Games, simulations, role-play*	1.77	1.23
Graphic organizers	1.71	1.50
Reflective writing/journaling	1.55	1.28

*$p < 0.05$ (t-test)

Means were calculated based on the following scale: *1* not used, *2* used once per semester, *3* used a few times a semester, *4* used most class sessions, and *5* used almost every class session

We cannot directly attribute the use of evidence-based approaches to FLC activities, as faculty members who use these approaches may be more likely to participate in FLCs. However, qualitative data collected from faculty members participating in FLCs suggest that the community encouraged and supported their adoption of these active learning strategies. As one community member commented, "I gain ideas that I can implement in my classes and share with colleagues" (Marbach-Ad et al., 2014, in press).

Student Reports of Teaching Practices They Experienced

Every semester, we conduct an exit survey of graduating seniors (see survey instrument at www.cmns-tlc.umd.edu/tlcmeasurementtools). The survey has many items, some of which probe the instructional and assessment practices students experienced in their undergraduate science coursework. Students rated the practices according to the following scale: 1 = none of my courses, 2 = rarely, 3 = sometimes—mostly in introductory courses, 4 = sometimes—mostly in upper-level courses, and 5 = in most courses. In analyzing these results, we often disaggregate the data by student major, gender, and whether the student completed all coursework at the university or transferred to the university after beginning their coursework at another institution.

The student exit survey asks about the same set of instructional and assessment practices as the STEP, which enables us to compare student reports to those of faculty members (e.g., Marbach-Ad et al., 2014, in press). However, we do so with caution. We ask faculty members to report on practices used within a single course, preferably a large enrollment course, while students are asked to report on their experiences across their science coursework as a whole. Even though the two data sets are not completely comparable, they provide a holistic picture of how evidence-based instructional and assessment approaches are being integrated into the undergraduate curriculum.

While we focus our efforts on a summative survey exploring the experiences of graduating seniors, it is also beneficial to collect information about students' experiences in a single course or a specific subset of courses. These more focused efforts may be particularly useful when evaluating the impact of a course redesign initiative or teaching and learning center intervention. Student reports can be collected via a variety of methods, including surveys (such as the exit survey), interviews, and focus groups. The latter options provide opportunities for students to elaborate on their responses and for the researcher to probe more deeply about specific aspects of the student experience.

Program Evaluation Using a Student Exit Survey

Our graduating senior exit survey provides valuable information on the extent to which different instructional practices are used in the classroom. Table 6.6 shows results from our 2011 exit survey. The most common instructional practices that students reported experiencing included extensive lecturing, communicating course goals and objectives, and answering questions from individual students during class sessions. We consider these all very traditional practices.

Students reported experiencing some practices more frequently in introductory courses than in upper-level courses. These included using personal response systems (i.e., clickers) and online modules that provide students with immediate feedback. These practices scale easily and so are appropriate for large enrollment courses. They also have the added benefits of helping instructors gauge student knowledge and providing differentiated support to students who come into the class with different backgrounds and knowledge. There were also practices that were more frequently used in upper-level courses, including working in groups outside of class time and interpreting graphical information. While these practices are certainly appropriate for upper-level courses, they also have merit for introductory classes; however, they may be more difficult to implement when class sizes are large. We continue to work with instructors to promote the use of these practices in introductory as well as advanced courses.

Students reported only rarely experiencing some recommended instructional approaches, including group work during class, out-of-class discussions, and reflective writing. Our informal communications with faculty members corroborate existing research that suggests that these approaches may be valued but can be

Table 6.6 Student reports of instructional techniques experienced in their undergraduate curriculum. Bolded values represent the modal answer

	None of my courses (%)	Rarely (%)	Sometimes, mostly intro courses (%)	Sometimes, mostly upper level courses (%)	In most courses (%)
Extensive lecturing	1	3	5	8	**83**
Communicating course goals	0	4	16	18	**63**
Answering questions in class	1	16	6	20	**58**
Asking students to interpret graphs	2	12	15	29	**42**
Multimedia instruction	2	24	21	19	**34**
Real-life problems	15	**37**	13	17	17
Class discussions	2	29	**32**	25	13
Graphic organizers	1	24	28	**35**	12
Group work outside of class	2	21	26	**41**	10
Personal Response System	4	12	**64**	10	10
Group work during class	5	**38**	30	18	8
Out-of-class discussions	11	**50**	24	10	5
Reflective writing/journaling	27	**43**	19	6	4
Debates in class	24	**43**	14	15	4
Online modules with immediate feedback	8	28	**55**	6	3
Games, simulations, role-play	33	**48**	12	4	2

Source: 2011 exit survey of graduating seniors from biology and chemistry (N = 285)

very difficult to implement (Taylor, 2011). These approaches can be particularly challenging to manage in large classes, and they are also challenging to grade. Our evaluation suggests that the TLC can play a vital role in supporting faculty members who seek to use these valuable but challenging to implement techniques.

Classroom Observations

Another way to assess the effective use of specific teaching practices is through direct observation of instructors in the classroom. Observations may be conducted by science education specialists, such as TLC staff, or by the instructor's peers. In our College, we employ both methods.

Upon faculty request, the TLC staff will conduct a general observation or one centered on a specific aspect of instruction. These observations tend to occur as a part of our consulting services (see Chap. 4). Faculty members request classroom observations for different reasons. For example, a faculty member who received negative feedback on end-of-course student evaluations may request a classroom observation to identify weaknesses and avenues for improvement. Faculty members seeking teaching-related grants, whether internally or externally funded, often request an observation to get feedback on the appropriateness of the planned intervention given the context of their course. If they receive the grant, the TLC can provide follow-up observations for comparison to the baseline observations, thereby facilitating evaluation of the teaching intervention.

Peer observations generally occur within a departmental structure. These observations can be informal, but increasingly our departments are adopting formal procedures. As described in Chap. 2, the TLC recently collaborated with the biology department in revising its peer review process to include all faculty members as both observers and observees. The goal was to increase the utility of peer reviews not just as a component of promotion and tenure decisions but also as a means of improving instruction. In some cases, faculty members request an observation outside of these formalized observations. Ad hoc observations allow observees to get feedback from peers who teach similar courses or who use methods that the observee would like to emulate.

Observations are most informative when they are guided by a detailed protocol. There are many validated classroom observation protocols, including

- Classroom Observation Protocol for Undergraduate STEM (COPUS): www.cwsei.ubc.ca/resources/COPUS.htm (Smith, Jones, Gilbert, & Wieman, 2013);
- Reformed Teaching Observation Protocol (RTOP): http://physicsed.buffalostate.edu/AZTEC/RTOP/RTOP_full/index.htm;
- Teaching Dimensions Observation Protocol (TDOP): tdop.wceruw.org/ (Hora & Ferrare, 2014); and
- The University of Minnesota: www1.umn.edu/ohr/teachlearn/resources/peer/instruments/.

Observation protocols vary in the degree to which they use selected response items (e.g., Likert-style items) and open-ended items. Selected response items are easier to quantify and have higher inter-observer reliability, while open-ended items allow for more nuanced characterization of classroom behavior.

The protocol chosen should reflect the purpose of the observation. If existing protocols do not adequately address this, they may be adapted, combined, and/or revised. Any protocols that are adapted or newly created should go through a validation process to ensure that they are accurately and reliably measuring the behaviors of interest. Prior to an observation, the observer should be familiar with the protocol. Depending on the complexity of the observation protocol that has been adopted, this preparation may range from a short training session to a comprehensive training program involving practice observations compared against those of an experienced observer.

Level 5: Impact

The TLC's overarching goal is to improve undergraduate instruction in our chemistry and biology departments. However, it is challenging to parse out the causes of observed improvements in undergraduate instruction. Changes in science instruction and student learning are relatively uncomplicated to measure, but attributing these changes to the actions of the TLC or to any of the many other institutional and national efforts to improve science education is difficult if not impossible. These efforts are strongly complementary, and their impacts are intertwined. Therefore, in this level of evaluation, we focus on impacts of very specific TLC-supported initiatives rather than the broader impact of the whole program. This targeted approach allows us to assess changes in areas where we would expect the TLC to have a measurable impact.

Aspects of impact that we measure include the following:

- Student satisfaction with their courses and course instruction;
- Student attitudes and beliefs about science;
- Student learning in individual courses; and
- The progression of student learning across sequences of related courses.

On the pages that follow, we elaborate on each of these components and provide examples from evaluation of the impacts of several initiatives supported by the TLC.

Student Satisfaction with Their Courses and Course Instruction

When we measure satisfaction, we measure variables including student interest, perceptions about the effectiveness of the course, and perceptions about the effectiveness of the instructor. Student interest in a course is influenced by the subject matter being taught as well as the way in which it is taught (Tobias, 1994), and high levels of interest can lead to increased student effort in a course (Schiefele, 1991). By increasing student satisfaction, we hope to increase student engagement with science, enhance their learning of scientific concepts and skills, and increase retention of students in STEM majors.

We measure student satisfaction with courses and instruction through end-of-course surveys that are required of all courses and administered through a central campus mechanism. We supplement this with course-specific supplemental surveys, focus groups, and interviews with individual students. Surveys and interviews can include broad questions about satisfaction with the course as a whole, as well as more specific questions related to different components of the course. Examples of these types of questions include:

- Was this instructor an effective teacher?
- Did the course satisfy your expectations?
- Would you recommend the course to others?
- Did the course prepare you for your intended career?
- Did you find the inquiry-based learning activities to be engaging?
- Did these activities help you better grasp course content?

Program Evaluation Using a Student Satisfaction Survey

A major impetus for creating the six-week GTA prep course (described in Chap. 5) was addressing undergraduate dissatisfaction with some laboratory and discussion sections that were customarily taught by GTAs. In an attempt to remedy this situation, beginning in Fall 2009, all incoming GTAs were required to take the newly created prep course that covered topics intended to help them become effective teachers. To evaluate the impact of the course for chemistry GTAs, we compared the end-of-course evaluations from the year prior to the creation of the course with the evaluations from the first two years in which all new GTAs took the course. The prep course was offered during the fall, so to measure the immediate impact of the prep course we used evaluations of their teaching from the following spring semester. These student evaluations of teaching showed that first-year GTAs who had completed the prep course in 2009 and 2010 received significantly higher scores on measures including effective teaching, being prepared for class, and showing respect for students than the previous cohort of new GTAs (who had not had the benefit of a prep course) (Fig. 6.6). Marbach-Ad et al. (2012) provide a more detailed description of the impact of the chemistry GTA prep course.

Student Attitudes and Beliefs About Science

Another way to measure the impact of professional development on undergraduate science instruction is to assess how science coursework influences students' attitudes and beliefs about science. For this aspect of impact, we often focus on

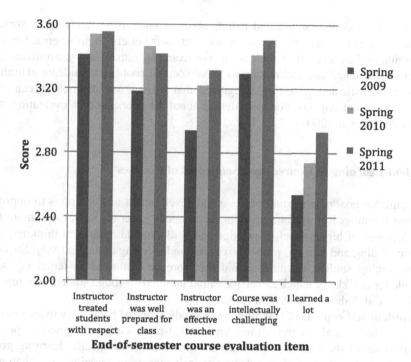

Fig. 6.6 Average course evaluation scores on select items for first-year GTAs who had not taken a teaching prep course (Spring 2009) and GTAs who had completed a prep course (Spring 2010 and 2011) (From Marbach-Ad et al. (2012). Reprinted with permission from the American Chemical Society)

non-science majors' attitudes and beliefs. These students may be required to take only a small number of courses in the natural sciences, so our goal in courses that enroll many non-science majors is to not only convey course-specific content but to also build awareness of science as a scholarly discipline and of what it means to think scientifically. These evaluations generally rely on validated tools that measure student views about science, such as the Views on Science-Technology-Society (VOSTS) (Aikenhead, Fleming, & Ryan, 1987; Aikenhead & Ryan, 1992).

Program Evaluation Using a Survey of Student Attitudes and Beliefs About Science

A microbiology course for non-science majors was revised to include pedagogical approaches intended to improve student understanding of science and to promote positive attitudes towards science. The revised course connected coursework to students' interests to demonstrate the relevance of science and employed many active learning techniques to engage students in scientific thinking. In addition to measuring student learning gains, we evaluated changes over the course of the semester in student attitudes and beliefs about science using the VOSTS survey. Student responses generally demonstrated that their beliefs about science became more nuanced as a result of the course. For example, rather than generalizing the benefits of science and technology to solve societal problems, students exhibited a more sophisticated view that recognized that science and technology can also have detrimental impacts. For more details about the course and its evaluation, see Marbach-Ad et al. (2009).

Student Learning in Courses and Sequences of Courses

The ultimate goal of providing professional development in teaching is to improve student learning. Reform initiatives in the sciences generally seek to promote the development of higher-level knowledge and skills, including critical thinking, scientific reading and writing, problem-solving, and working collaboratively. Success in improving student learning should therefore be evaluated in terms of these knowledge and skills, which can be measured indirectly through student self-reports and/or through direct assessments of learning.

Student self-reports of learning have the advantage of being easy to measure in terms of student and researcher time. Student self-reports can be targeted to specific components of the course (e.g., a lab activity) as well as specific learning goals (e.g., improving writing skills). However, self-reports of learning are inherently subjective and may not reflect learning gains as measured through direct assessment of knowledge and skills. Validated instruments to collect student self-reports of learning include:

- Student Assessment of their Learning Gains (SALG; www.salgsite.org/);
- Classroom Undergraduate Research Experience survey (CURE; www.grinnell. edu/node/25703); and
- Survey of Undergraduate Research Experiences (SURE; www.grinnell.edu/ academics/areas/psychology/assessnebts/sure-iii-survey.

Direct assessment of knowledge and skills can occur through course exams, concept inventories, performance assessments, portfolios, and standardized assessments. These direct assessments should be targeted at the specific skills and knowledge that the intervention seeks to improve. Direct assessments can measure student learning gains in a single course, across a sequence of courses, or both, depending on the nature and scope of the intervention. In some cases, it is best to use a self-developed assessment that has been created specifically for that particular context. In other cases, self-developed assessments can be supplemented with or replaced by existing validated instruments such as those listed below.

- General assessments

  - Collegiate Learning Assessment (CLA): used and discussed in *Academically Adrift* (Arum & Roksa, 2011)
  - Critical Thinking Assessment Test (CAT): www.tntech.edu/cat

- Domain-specific concept inventories
  - Force Concept Inventory (Hestenes, Wells, & Swackhamer, 1992)
  - Host Pathogen Interaction Concept Inventory (Marbach-Ad, McAdams, et al., 2010)
  - Conceptual Inventory of Natural Selection (CINS) (Anderson, Fisher, & Norman, 2002)
  - Genetics Concept Assessment (Duit & Treagust, 2003; Smith, Wood, & Knight, 2008)

Program Evaluation Using Direct Assessment of Student Learning

Direct assessments of student learning gains generally center on a single course that has undergone major revision. One such example is that of an immunology lab course that was redesigned to focus on improving students' skills in scientific reading and writing through authentic research. The activity was structured in three successive stages, each of which included several labs and related instruction. In Stage 1, students were familiarized with the basic components of a research paper. In Stage 2, students learned specific skills required for writing research papers. In Stage 3, students applied these skills as they conducted and wrote about their own research.

To measure student success in improving their scientific writing skills, the course instructors analyzed writing assignments according to a set of rubrics. A comparison of student writing from Stages 1 and 3 demonstrated significant gains over time in the quality of student writing for two of the three major components of research papers: the introduction to the study and the discussion of research findings (Figure 6.7). There was a non-significant trend for improvement in writing up the methods and results section of their research paper. See Senkevitch et al. (2011) for a more detailed description of the intervention and its assessment.

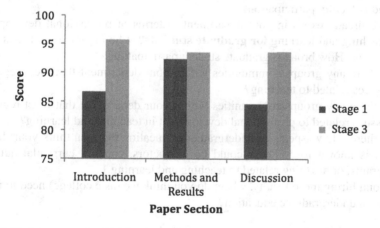

Fig. 6.7 Student scores on initial (Stage 1) and final (Stage 3) writing assignments by component of research paper (Adapted from Senkevitch et al. (2011), with permission from the Journal of Microbiology and Biology Education)

Conclusion

Evaluation plays an obvious and important role in any professional development program, but conducting robust and useful evaluation is far from simple. Among the many inherent challenges are developing and validating appropriate assessment instruments, the need for working with institutional human subjects review boards to ensure the welfare of those from whom data is collected, and the time required to analyze and interpret thoroughly the resultant data. We have found it valuable to employ a variety of evaluation strategies to reach a comprehensive understanding of the quality and impact of professional development initiatives.

When done well, evaluation provides program developers with feedback essential to the creation and refinement of professional development programs. Evaluation also provides valuable data for current and potential funders of educational initiatives. Finally, the results of thoughtful evaluation, when disseminated through professional presentations and papers, contribute to the growing body of knowledge on improving science education.

Implementation Guide

Guiding Questions for Departmental Meetings—Needs Assessment

We intend to use these questions to stimulate our discussions. We are distributing them ahead of the meeting in case you want to discuss them with colleagues.

1. What already exists in your department in terms of professional development in teaching and learning **for faculty**? Who organizes and runs these activities? How broad is faculty participation?
2. What already exists in your department in terms of professional development in teaching and learning **for graduate students**? Who organizes and runs these activities? How broad is graduate student participation?
3. Are there any groups/communities within your department that meet regularly on issues related to teaching?
4. Are there any groups/communities within your department that meet regularly on issues related to professional development in teaching and learning?
5. Are there a few aspects of undergraduate education that you think your department is known for? (These could be instructors, courses, curricula, outreach programs, or anything related to teaching and learning.)
6. Beyond hiring more faculty, where do you think we (as a college) need to invest more in undergraduate education?

Sample Completed Faculty Learning Community Observation Protocol

Name of the community: Host Pathogen Interaction (HPI)

Date of community meeting: March 16, 2011

Number of participants: 15

Physical environment (room, refreshments, etc.): Very conducive to group discussion as all participants visible and included; available computer facilities; back table convenient for light lunch/refreshments; and all participants seemed comfortable with setting.

Length of meeting: Flexible beginning for lunch; "formal focus" from 12:30 to 2:05

Who are the people in the community in terms of gender, position, disciplines, age, and experience? Participants included 5 females, including the Chair of the meeting, and 9 males. The HPI group included a rich mix of faculty in terms of content/course responsibilities, rank, age, experience, and research and teaching interests.

Opening remarks (Who leads the meeting)? The Chair of the meeting is the leader of the FLC, and served as PI for multiple grants that supported the HPI FLC. The meeting began with enthusiasm and air of "seamless continuation of valuable discourse" about enhancing student learning and the framework of student learning.

Are there multiple people contributing? At this meeting, 100 % of the participants contributed to the discussion. One person did not make a comment until 1:25 PM and one person arrived late, but by 1:40 PM, all HPI members present had made substantive comments, asked questions, and/or initiated new topics.

Is the community forming subgroups? No, the group was focused and cohesive and only one short incident of "sidebar" commenting to colleagues was observed.

Does the community have a goal? Yes. To the External Evaluator, the goal of meeting as a community seems to be for faculty to collaboratively examine, share, and implement content connections, teaching strategies, classroom practices, and curriculum changes in HPI program that can/will enhance student learning, long-term retention, and success.

Is the community working towards meeting their goal(s)? Yes. There was an extensive discussion of recent and potential changes in an introductory microbiology course that was recently divided into separate courses for majors and non-majors. The discussion focused on preliminary student data related to differences in requirements for prerequisite courses in genetics for majors and non-majors, which seems to impact students' success and appropriate teaching strategies.

Is there discussion of classroom experience and student learning? Yes. There was a discussion of what concept mapping is, and faculty members shared positive and negative experiences with concept mapping in their HPI classes. Time was spent discussing lab skills expectations of students and how to help grad students anticipate these in various courses. The group continued to support/value by shared examples the use of "anchor organisms" in labs and "unifying concepts" across all courses in the major.

Are community members cooperating with each other? Yes, see previous comments. It is also important to note that the group "leader" is truly an expert facilitator.

What are they focusing on in terms of professional development: curriculum change, pedagogical change, assessment, etc.? All of the above. References were given for finding technologies, using lab equipment, assessing student learning in lecture and lab, and finding resources for internal (UMD) or external grant opportunities.

Is there a sense of acceptance to share ideas in the group? Yes, as described earlier, and evidenced in only one person leaving the meeting before closing.

Does the community use outside experts? At this meeting, the community only used printed materials, such as a paper about the use of concept mapping in higher education.

Is there discussion about future projects, grants, research, etc.? Yes. The discussion included a request for student data for a future research study, reading on widespread biological sciences curricular reform to inform future decisions about courses for majors and non-majors, and choices of topics and speakers for future meetings.

Is there continuity (planning for the next meeting, building off of the last meeting)? Yes. This learning community has a rich set of past experiences and demonstrates both a collaborative, continued momentum and effective resolve at this time.

References

Aikenhead, G. S., Fleming, R. W., & Ryan, A. G. (1987). High-school graduates beliefs about science-technology-society. 1. Methods and issues in monitoring student views. *Science Education, 71*(2), 145–161.

Aikenhead, G. S., & Ryan, A. G. (1992). The development of a new instrument—Views on science-technology-society (Vosts). *Science Education, 76*(5), 477–491.

American Association for the Advancement of Science (AAAS). (1990). *Science for all Americans.* New York: American Association for the Advancement of Science.

American Association for the Advancement of Science (AAAS). (2011). *Vision and change: A call to action.* Washington, DC: AAAS.

Anderson, D. L., Fisher, K. M., & Norman, G. J. (2002). Development and evaluation of the conceptual inventory of natural selection. *Journal of Research in Science Teaching, 39*(10), 952–978. doi:10.1002/Tea.10053.

Arum, R., & Roksa, J. (2011). *Academically adrift: Limited learning on college campuses.* Chicago: University of Chicago Press.

Austin, A. E., & Sorcinelli, M. D. (2013). The future of faculty development: Where are we going? *New Directions for Teaching and Learning, 133*, 85–97.

Austin, A. E., Sorcinelli, M. D., & McDaniels, M. (2007). Understanding new faculty: Background, aspirations, challenges, and growth. In R. Perry & J. Smart (Eds.), *The scholarship of teaching and learning in higher education: An evidence-based perspective* (pp. 39–89). Dordrecht, The Netherlands: Springer.

Boyer Commission on Undergraduates in the Research University. (1998). *Reinventing undergraduate education: A blueprint for America–s research universities.* Stony Brook, NY: State University of New York at Stony Brook.

Boyle, M. A., & Crosby, R. (1997). Academic program evaluation: Lessons from business and industry. *Journal of Industrial Teacher Education, 34*(3), 81–85.

Chen, W., Kelley, B., & Haggar, F. (2013). Assessing faculty development programs: Outcome-based evaluation. *Journal Centers for Teaching and Learning, 5,* 107–119.

Colbeck, C. L. (2003). *Measures of success: An evaluator's perspective.* Paper presented at the CIRTL Forum, Center for the Integration of Research, Teaching and Learning, Madison, WI.

Duit, R., & Treagust, D. F. (2003). Conceptual change: A powerful framework for improving science teaching and learning. *International Journal of Science Education, 25*(6), 671–688. doi:10.1080/09500690320000076652.

Fink, L. (2013). Innovative ways of assessing faculty development. *New Directions for Teaching and Learning, 2013*(133), 47–59.

Gibbs, G., Knapper, C., & Piccinin, S. (2008). Disciplinary and contextually appropriate approaches to leadership of teaching in research-intensive academic departments in higher education. *Higher Education Quarterly, 62*(4), 416–436.

Guskey, T. R. (2000). *Evaluating professional development.* Thousand Oaks, CA: Corwin.

Handelsman, J., Ebert-May, D., Beichner, R., Bruns, P., Chang, A., DeHaan, R., . . . Wood, W. B. (2004). Scientific teaching. *Science, 304*(5670), 521–522.

Henderson, C., Beach, A., & Finkelstein, N. (2011). Facilitating change in undergraduate STEM instructional practices: An analytic review of the literature. *Journal of Research in Science Teaching, 48*(8), 952–984.

Hestenes, D., Wells, M., & Swackhamer, G. (1992). Force concept inventory. *The Physics Teacher, 30,* 141–158.

Hines, S. R. (2009). Investigating faculty development program assessment practices: What's being done and how can it be improved. *Journal of Faculty Development, 23*(3), 5–19.

Hora, M., & Ferrare, J. (2014). *The teaching dimensions observation protocol (TDOP) 2.0.* Madison, WI: University of Wisconsin-Madison, Wisconsin Center for Education Research.

Hurtado, S., Eagan, K., Pryor, J. H., Whang, H., & Tran, S. (2012). *Undergraduate teaching faculty: The 2010–2011 HERI faculty survey.* Los Angeles, CA: Higher Education Research Institute.

Kirkpatrick, D. L. (1994). *Evaluating training programs: The four levels.* San Francisco: Berrett-Koehler.

Kirkpatrick, D. L. (1998). *Evaluating training programs: The four levels* (2nd ed.). San Francisco, CA: Berrett-Koehler.

Kucsera, J. V., & Svinicki, M. (2010). Rigorous evaluations of faculty development programs. *Journal of Faculty Development, 24*(2), 5–18.

Levinson-Rose, J., & Menges, R. J. (1981). Improving college teaching: A critical review of research. *Review of Educational Research, 51*(3), 403–434.

Marbach-Ad, G., McAdams, K., Benson, S., Briken, V., Cathcart, L., Chase, M., . . . Smith, A. (2010). A model for using a concept inventory as a tool for students' assessment and faculty professional development. *CBE Life Science Education, 9,* 408–436.

Marbach-Ad, G., McGinnis, J. R., Pease, R., Dai, A., Schalk, K. A., & Benson, S. (2009). Promoting science for all by way of student interest in an undergraduate microbiology course for non-majors. *Journal of Microbiology & Biology Education, 10,* 58–67.

Marbach-Ad, G., Schaefer, K. L., Kumi, B. C., Friedman, L. A., Thompson, K. V., & Doyle, M. P. (2012). Development and evaluation of a prep course for chemistry graduate teaching assistants at a research university. *Journal of Chemical Education, 89*(7), 865–872.

Marbach-Ad, G., Schaefer, K. L., & Thompson, K. V. (2012). Faculty teaching philosophies, reported practices, and concerns inform the design of professional development activities of a disciplinary teaching and learning center. *Journal on Centers for Teaching and Learning, 4*, 119–137.

Marbach-Ad, G., Schaefer Ziemer, K. L., Thompson, K. V., & Orgler, M. (2013). New instructor teaching experience in a research-intensive university. *Journal on Centers for Teaching and Learning, 5*, 49–90.

Marbach-Ad, G., Shields, P. A., Kent, B. W., Higgins, B., & Thompson, K. V. (2010). Team teaching of a prep course for graduate teaching assistants. *Studies in Graduate and Professional Students Development, 13*, 44–58.

Marbach-Ad, G., Shaefer-Ziemer, K., Orgler, M., & Thompson, K. (2014, in press). Science teaching beliefs and reported approaches within a research university: Perspectives from faculty, graduate students, and undergraduates. *International Journal of Teaching and Learning in Higher Education, 26*(2).

NRC. (2003). *National Research Council, Bio 2010: Transforming undergraduate education for future research biologists*. Washington, DC: National Academy Press.

National Science Foundation (NSF). (1996). *Shaping the future: New expectations for undergraduate education in science, mathematics, engineering, and technology*. Washington, DC: National Science Foundation (NSF).

Plank, K. M., & Kalish, A. (2010). Program assessment for faculty development. In K. J. Gillespie & D. L. Robertson (Eds.), *A guide to faculty development* (2nd ed., pp. 135–150). Hoboken, NJ: Wiley.

President's Council of Advisors on Science and Technology (PCAST). (2012). Engage to excel: Producing one million additional college graduates with degrees in science, technology, engineering, and mathematics. Available at www.whitehouse.gov/sites/default/files/microsites/ostp/pcast-engage-to-excel-final_2-25-12.pdf

Quardokus, K., & Henderson, C. (2014). *Using department-level social networks to inform instructional change initiatives*. Paper presented at the National Association for Research in Science Teaching (NARST) annual meeting, Pittsburg, PA.

Robertson, D. L. (2010). Establishing an educational development program. In K. J. Gillespie & D. L. Robertson (Eds.), *A guide to faculty development* (2nd ed., pp. 35–52). Hoboken, NJ: Wiley.

Schiefele, U. (1991). Interest, learning, and motivation. *Educational Psychologist, 26*(3–4), 299–323.

Senkevitch, E., Marbach-Ad, G., Smith, A. C., & Song, S. (2011). Using primary literature to engage student learning in scientific research and writing. *Journal of Microbiology and Biology Education, 12*, 144–151.

Smith, M. K., Jones, F. H., Gilbert, S. L., & Wieman, C. E. (2013). The classroom observation protocol for undergraduate STEM (COPUS): A new instrument to characterize university STEM classroom practices. *CBE Life Sciences Education, 12*(4), 618–627. doi:10.1187/cbe. 13-08-0154.

Smith, M. K., Wood, W. B., & Knight, J. K. (2008). The genetics concept assessment: A new concept inventory for gauging student understanding of genetics. *CBE Life Science Education, 7*(4), 422–430.

Sorcinelli, M. D. (2002). Ten principles of good practice in creating and sustaining teaching and learning centers. In K. H. Gillespie, L. R. Hilsen, & E. C. Wadsworth (Eds.), *A guide to faculty development: Practical advice, examples, and resources* (pp. 9–23). Bolton, MA: Anker.

Sorcinelli, M. D., Austin, A. E., Addy, P. L., & Beach, A. L. (2006). *Creating the future of faculty development: Learning from the past, understanding the present*. Bolton, MA: Anker.

Stufflebeam, D. L., Madaus, G. F., & Kellaghan, T. (2000). *Evaluation models: Viewpoints on educational and human services evaluation* (Vol. 49). Berlin, Germany/New York, NY: Springer.

Taylor, A. (2011). Top 10 reasons students dislike working in groups … and why I do it anyway. *Biochemistry and Molecular Biology Education, 39 (2), 219–220, 39*(2), 219–220.

Tobias, S. (1994). Interest, prior knowledge and learning. *Review of Educational Research, 64*(1), 37–54.

Travis, J. E., Hursh, D., Lankewicz, G., & Tang, L. (1996). Monitoring the pulse of the faculty: Needs assessment in faculty development programs. In L. Richlin (Ed.), *To improve the academy* (Vol. 15, pp. 95–113). Stillwater, OK: New Forums Press.

University of Maryland Office of the Provost. (2011). In Mission and Goals Statement University of Maryland, College Park. Retrieved from http://www.provost.umd.edu/Documents/UMCP-Mission-Statement-Final-2011.pdf. (Ed.).

Wieman, C. (2007). Why not try a scientific approach to science education? *Change.*http://www.changemag.org/Archives/Back%20Issues/September-October%202007/index.html

Wieman, C., Perkins, K., & Gilbert, S. (2010). Transforming science education at large research universities: A case study in progress. *Change.*http://www.changemag.org/Archives/Back%20Issues/March-April%202010/transforming-science-full.html

Wright, M. C. (2011). Measuring a teaching center's effectiveness measuring a teaching center's effectiveness. In C. E. Cook & M. Kaplan (Eds.), *Advancing the culture of teaching on campus: How a teaching center can make a difference* (pp. 38–49). Sterling, VA: Stylus.

Chapter 7
Concluding Thoughts

In this book, we describe a 10-year effort to improve undergraduate science instruction through professional development in teaching. This effort differs from many similar initiatives nationwide in that our guiding philosophy is that professional development should be strongly rooted in the relevant scholarly disciplines of the faculty and graduate students being served. Our Teaching and Learning Center (TLC), which serves the biology and chemistry departments at the University of Maryland, predicates its programming on pedagogical content knowledge (PCK) that addresses the specific issues of teaching the sciences. This disciplinary focus is strengthened and complemented by collaboration with university-wide faculty development specialists, such as those at the Center for Teaching Excellence, who offer interdisciplinary expertise and programming.

Our work within the biology and chemistry departments we serve has afforded us a strong understanding of our constituent community. We recognize that this community is diverse. At a research-intensive university such as ours, science departments tend to be large and include research and instructional faculty as well as graduate students, many of whom take part in providing undergraduate instruction and some of whom will comprise the faculty of tomorrow. We are sensitive to the different needs, interests, and professional responsibilities of these different populations.

A growing body of literature recognizes the need for professional development in teaching for graduate students, but such opportunities are scarce. We have created programming tailored to the developmental trajectory of graduate students, beginning with a mandatory training course for all incoming graduate teaching assistants and continuing with optional programs for graduate students who wish to develop greater expertise in teaching and learning. The TLC also seeks to foster their acculturation as science educators and communicators.

Our research into the background and training of new faculty members suggests that, while they come with different levels of experience and expertise, most benefit from comprehensive professional development in teaching. As with graduate

© Springer International Publishing Switzerland 2015
G. Marbach-Ad et al., *A Discipline-Based Teaching and Learning Center*,
DOI 10.1007/978-3-319-01652-8_7

students, we offer new faculty varied professional development opportunities that include both required and optional components. We introduce them to topics in teaching and learning through welcome workshops, provide salient science education literature, and encourage them to take advantage of the TLC's services as needed.

Teaching science is a complex task, and the science education community is continually adding to the body of knowledge on effective teaching approaches. Our understanding of how to prepare scientifically literate graduates and the next generation of scientists is dynamic. Even experienced faculty members benefit from ongoing professional development as they hone their teaching expertise and adapt their teaching to reflect emerging research. Our professional development activities for all faculty members include seminars and workshops that are designed for broad audiences, as well as highly individualized consulting services. These consultation services support individuals and groups, including several faculty learning communities (FLCs).

The TLC emphasizes research and evaluation in all aspects of its work. Our initial needs assessment shaped the activities described in the preceding chapters. Our multi-level, ongoing program evaluation informs the enhancement of these activities and the development of future programming. Through research, evaluation, and reflection, we have continually refined the TLC and made it a valued resource for our stakeholders.

Lessons Learned

The TLC that exists today is the culmination of a 10-year effort to build on and enhance existing professional development. Along the way, we have learned many lessons that reflect the growth process of the TLC, the challenges we faced, and the successes we achieved, as well as the wisdom of the TLC's many collaborators. The TLC's development has been organic and iterative. Below, we highlight key lessons that we believe could be helpful for other universities that seek to implement discipline-based professional development programs.

Pursue both top-down and bottom-up changes: We found that there is no single way to foster change, and that successful change initiatives require the support and engagement of diverse stakeholders. The TLC was one component of a top-down change initiative in which college and departmental leadership sought to change the culture around teaching and learning. The College initiated, supported, and emphasized the importance of the Center and its activities, which lent the TLC credibility as well as sustainability. The TLC was also part of bottom-up change initiatives that were led by faculty members and graduate students who were deeply committed to improving their own teaching. These individuals served as key change agents and drew their peers to the Center and its activities. One of the TLC's roles in these initiatives was connecting college and departmental leadership with the current and future faculty members who shared a vision of improving instruction in the sciences.

Recognize the centrality of the department: We view the department as the functional unit of the university in terms of enhancing teaching and learning. Embedding a teaching and learning center within the department(s) it serves, and integrating its activities into departmental activities, can enhance the effectiveness of professional development. This specialization will also attract more participants, particularly when its activities are imbued with PCK.

Involve key departmental personnel: We previously mentioned the importance of the involvement of faculty members and graduate students engaged in improving their teaching. We also encourage the involvement of discipline-based education researchers (DBERs), who are in a unique position to solidify the connections between disciplinary norms, content, and pedagogy. Additionally, a disciplinary teaching and learning center should involve individuals who hold status among their peers. Their involvement lends credibility to the center and its mission. They can serve in formal roles, such as on a teaching and learning center advisory board, and should be encouraged to play key supporting roles in center activities.

Acculturate new instructors: One of the TLC's biggest successes has been its programming for incoming faculty members and graduate students. New faculty members and graduate students undergo a process of socialization as they assume or prepare to assume faculty roles. Many of them have received extensive training in research, while their training in teaching has been more limited. We provide training and resources to new faculty members and new graduate teaching assistants (GTAs), which opens the door for those who seek additional professional development.

Tailor professional development programming to the diverse population served: The TLC serves a broad population that includes instructional faculty, science researchers, science education researchers, and graduate students. Although these populations share a disciplinary focus, their different roles result in different needs. A teaching and learning center can offer multiple types of activities to reflect these needs, and this programming should be sensitive to each population's priorities, professional strengths, and professional weaknesses. We also find it valuable to recognize that individuals vary in their level of interest in pursuing professional development in teaching.

Encourage collaborations around teaching and learning: Much of the TLC's success rests in its ability to harness the power of groups of collaborators. We have strongly encouraged collaboration through faculty learning communities (FLCs). Groups in general, and formal FLCs in particular, can have greater capacity to scale up change or take on large-scale change efforts than can most individuals working alone. Additionally, well-established FLCs are in a better position to secure grant funding and other resources than individuals or ad hoc groups of faculty. In order to enhance their potential, communities need support, guidance, and management. Discipline-based teaching and learning centers are well positioned to provide this guidance in a manner that reflects best practices in science education.

Evaluate, evaluate, evaluate: Change initiatives must be built on a foundation of evidence. Through a strong focus on evaluation, a teaching and learning center can demonstrate that science education research can—and should—be as rigorous as scientific research. This is particularly critical for gaining faculty buy-in within research-intensive universities, where research evidence constitutes the common language. Deeply integrating analytics and research into change initiatives establishes the credibility of both the initiatives and their advocates. Furthermore, well-conceived and documented evaluation can serve as a powerful impetus for additional reform.

In this book, we have presented a discipline-based teaching and learning center as an effective model for professional development. We have shared the theory underpinning this model and offered suggestions for implementing it at other institutions. This model was formed not only from our own work and experience, but also from the expertise of our many colleagues who share our commitment to improving undergraduate science education. We hope that this book serves as an inspiration and guide to others to continue in this work, and we look forward to learning from their future endeavors.